THE ORIGIN
OF
LIFE

GOD, DARWIN AND CHANCE

FERNANDO ORREGO
Origin of Living Beings Program
Universidad de los Andes
Santiago, Chile

ISBN: 1494291975
ISBN 13: 9781494291976
Library of Congress Control Number: 2014906175
CreateSpace Independent Publishing Platform, North Charleston, South Carolina

Dedicated to all who made me a scientist:
My parents
Lucy Porter
My teachers at Belgrano Day School (Buenos Aires)
Roque Esteban Scarpa
Armando Roa
Hermann Niemeyer
Fritz Lipmann
To all the many bright students who have worked with me,
And, especially, to María Cristina Sánchez, my wife who, with her courage and inge-
nuity, allowed me to remain in the paths of science.

ACKNOWLEDGEMENTS

I am grateful to M. Valentina Valdés, Victor Pons, Katerina Ferrat and Ursula Wyneken for their help with the text, and to Mauricio Sandoval, Carlos Lagos and Catherine Ocqueteau for their fine illustrations and, especially, to M. Carolina Barros for her beautiful cover.

INDEX

REFERENCES

FIGURE LEGENDS

PREFACE

The appearance on our planet of the first living cells was the consequence of a large series of apparently unconnected events that started at the beginning of our Universe. Starting, however, with the formation of our Solar System, we shall follow the origins of some of the planetary conditions for life, as well as the origin of organic and inorganic substances, and their assembly, which allowed the first independent cell to appear. For this, the Darwinian principles of variation, natural selection, homology and common ancestry, that have a much broader scope than that of biology, shall be used whenever possible.

When studying this problem I realized that it could only be understood by placing the very large number of facts that have been discovered in recent decades, into a proper perspective. This involves some knowledge, not only of Biology, but also of Chemistry, Physics, including Astrophysics, Cosmology, Geology and even Natural Philosophy, fields in many of which I have only a grasping of their basic principles.

I have tried to keep the language in this book as clear and non-technical as possible, leaving the more technical or specialized as footnotes. However, this is not a book in which science is made popular for the sake of the "general public", but a rigorous scientific endeavour, with which I intend to answer, in part, what Charles Darwin called "that great fact –that mystery of mysteries- the first appearance of new beings on this Earth"[1], but that now appears possible, at least in part, thanks to the large number of high quality scientific research produced since the mid-twentieth century, some of which the interested reader

---

1. C. Darwin in "The Voyage of the Beagle", October 8th, 1835

may find in the excellent "The Origins of Life", edited by D. Deamer and J.W. Szostak[2].

The purpose of this book, however, is not to review the scientific literature on this subject but rather, it is an effort to reconstruct the extraordinary history of events that led to the emergence of the first cell on our planet.

2. D. Deamer, J.W. Szostak (Eds.) The Origins of Life. Cold Spring Harbor Laboratory Press, N.Y. (2010).

PART ONE

PATHS TOWARDS LIFE

CHAPTER 1

FROM COSMIC DUST TO OUR SOLAR SYSTEM

The solar system in which we live, is thought to have been born some 4570 million years ago, in a large, cold, dark molecular cloud, formed by gas and by tiny granules –the cosmic dust– in one of the spiral arms of our galaxy, the Milky Way[1].

This molecular cloud, in the region where our protosun was to be born, had a temperature of about 10 K, which is close to absolute zero (0 K, -273.15° C)[2], and was relatively dense -it was a molecular core-, compared to other regions of the cloud. This cold and darkness was a consequence of the presence of cosmic dust, especially dark carbon grains, that prevented the intense electromagnetic radiation present in the interstellar medium from penetrating the cloud. This radiation would have destroyed any incipient structure in the protosolar system. Molecular clouds owe that name because most of the 150 known interstellar organic molecules are present in them and, also, because they were formed inside them as well as throughout the Universe by intense chemical reactions[3-5].

Interstellar clouds are made up of gas and dust particles. In low density diffuse clouds –the precursors of molecular clouds– the gas is composed of 89% hydrogen, both atomic (H) and molecular (H_2), helium (9%) and heavier elements (2%). The two former were generated, together with some lithium, in the very early universe, some 2 or 3 min after the Big Bang. The rest of the gas components are water (H_2O), methane (CH_4), ammonia (NH_3), oxygen (O_2), nitrogen (N_2), carbon monoxide (CO), carbon dioxide (CO_2), cyanide-containing compounds (- CN), methanol (CH_3OH), formaldehyde (H_2CO), ethanol (CH_3CH_2OH), formic acid (HCOOH), carbon-oxysulfide (COS) and a host of other minor components[3,5,7]. In dense molecular clouds, however, due to their very low temperature, especially in the denser molecular cores that are below the freezing point of most of the gas components (Table 1), such gases accreted as ices on the surface of the more abundant granules (the core-mantle grains), possibly in a layered order, with the more easily frozen (i.e. those with a higher freezing point) accreting first, closer to the core surface. H, He and neon (Ne), because of their very low freezing temperature, however, remained in the gas phase[6].

Figure 1: THE SNAKE DARK CLOUD
A very large dark, cold molecular cloud, formed by cosmic dust and gases, where stars, including solar systems, are born, is shown against a background of stars. (Credit: ESA).

TABLE 1
FREEZING TEMPERATURE OF GASES (°C)

H_2O	0
CO_2	-78
NH_3	-78
CO	-205
Nitrogen	-210
Oxygen	-218.8
Neon	-248.7
Hydrogen	-259.1
Helium	-272.2

COSMIC DUST GRANULES

Three main types of dust grains are generally accepted to exist inside molecular clouds. The most abundant, and larger (about 0.2 by 0.4 μm.), the core-mantle grains, constitute 80% of all the interstellar dust mass[8,9]. They consist of a rocky core, of about 0.05 μm in diameter, formed by a large number of nanoparticles, that contain all the heavier elements that shall, ultimately, give rise in our solar system to the rocky planets, to the cores of the large gaseous planets and to asteroids. These elements are C, Si, Mg, Al, Ca, Fe, Ni, and others[6,8].

But the grain cores are far from being a homogeneous population and, for example, their Fe content varies between 14% and 43% of the dust particle mass[3]. This indicates that for our Earth to have formed in its present state, a unique mix of cosmic grain was needed.

The origin of these grains is multiple: most seem to derive from stars that reached the red giant phase, both carbon-rich and oxygen and Si-rich stars, whose outer parts cooled, solidified and fragmented, giving rise to grain cores. Supernovae remnants are also rich sources of interstellar grains [10,11].

Grains are further processed by supernovae shocks, by UV radiation and by collision with other grains. This secondary processing is thought to give rise to the other two main types of cosmic grains: the carbonaceous one, that measure about 0.005 μm and that contain carbon, as graphite, and those that contain a mixture of polycyclic aromatic hydrocarbons (PAHs), of about 0.002 μm in size. Both of these small grains constitute about 10% each of the interstellar dust mass [6-8].

As mentioned, the core-mantle grains, because of their low temperature, accreted onto their core most of the gaseous species present in their vicinity, forming the grain mantle, 60% of which may be frozen water. On their surface, because of photochemical reactions, a coating of undefined organic material was also present. Core-mantle grains are also a place of considerable chemical activity in which other compounds, including formaldehyde, that shall later give rise to sugars, are formed. It is thought that atomic hydrogen (H) suffers a catalytic process on the grain surface, by which it binds to another H atom forming H_2, molecular hydrogen, a fundamental constituent of interstellar space[6,7].

One of the most important functions of these dust grains is to absorb or scatter radiation, especially visible light and the more energetic UV radiation that could destroy the formation of solar systems like our own, as has been the case many times. This action allowed the formation, in molecular cloud cores, of extremely cold and dark regions -true solar system wombs- where gas and dust could now form a sun-like star and its associated protoplanetary disk. Such womb regions could not have been formed in the early universe, where no dust was yet present, and where frequent supernovae explosions and huge radiation and gravitational fields would disrupt any fragile embryonic solar system.

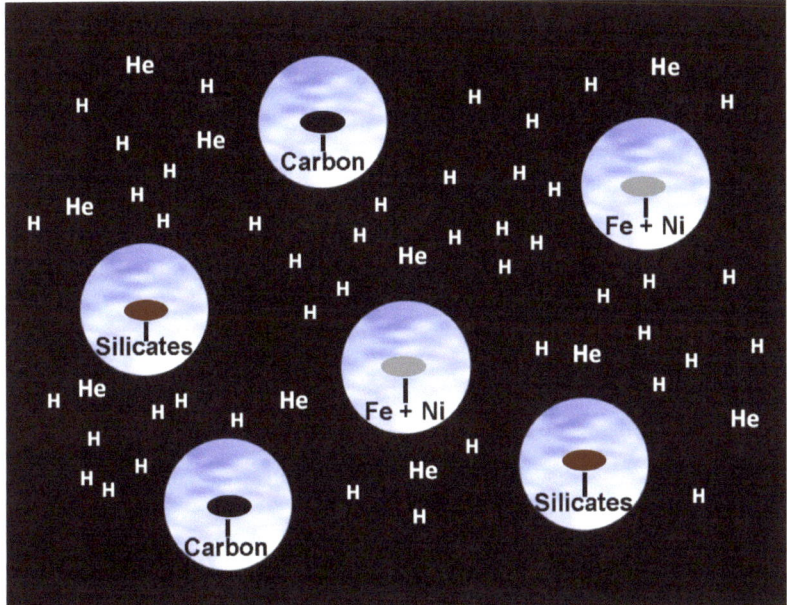

Figure 2: CORE-MANTLE COSMIC GRAINS
Grains with their cores of different composition (e.g. iron-nickel; carbon or silicates) and their mantle of frozen gas (H_2O, CO, NH_3, N, NH_3, CH_3OH, CH_2O, etc) are shown. Their diameter is 0.4 μm and they constitute about 80% of all dust grains. H and He, that do not freeze, are shown outside the grains. (Credit: M. Sandoval).

BEGINNINGS OF OUR SOLAR SYSTEM

It is thought that the explosion of a supernova near "our" molecular cloud, that generated a strong shock wave, induced a pressure increase in one or more of the molecular cores. This produced a series of events leading to our solar system and, also, to other neighboring ones, giving rise to a star cluster, whose stars later drifted away. This is supported by the abundance of bright blue stars, that end as supernovae, in the same spiral arms of the Milky Way that are also rich in molecular clouds. The pressure increase in the molecular core, by elevating the density of gas and dust, also increased their gravitational inward pull, starting a chain reaction in which the increment in the gravitational field attracted more gas and dust, originating a positive feedback -one of the most powerful mechanisms in nature- until a substantial mass of gas and dust accumulated. Prior to this, molecular cloud cores were in equilibrium, in which the gravitational attraction of gas and dust tended to compact the cloud, but this was balanced by magnetic pressure and other mechanisms, that tended to expand it. The increase in pressure induced by the supernova shock broke this equilibrium, leading to the gravitational field to prevail and to the collapse of the cloud. This process is now known to be rather turbulent, in which the pressure shock did not act on every point of the core with equal force, inducing a rotational spin of the compacting core[12,13].

In the contracting and rotating cloud, most of the mass of the original core (99.87%) accumulated in the central region, where it would give rise to a protosun[14]. In it, because of the gravitational pull, pressure and temperature increased progressively, until temperature reached some 5 million K or more and hydrogen nuclei started to fuse among themselves forming helium, together with the release of large amounts of energy by the conversion of mass into energy, following the well known Einstein relation: energy released equals the loss of mass multiplied by the velocity of light squared ($E=mc^2$) [note 2].

The rest (0.13%) of the dust and gas originally present in the system, formed a rotating protoplanetary disk. At first, when the mass accumulated in the center was not large, because of the conservation of angular momentum, the disk rotated slowly and the temperature remained low, so that the core-mantle cosmic grains contacted each other also at slow speed, without much increase in temperature. This, because of restricted melting of the frozen gases on grain surfaces, followed by fusion with the

---

Note [2]: In the H fusion process, four H nuclei (mass 4.032 atomic mass unit, a.m.u) fuse to form one He nucleus (mass 4.004 a.m.u) so that in the process there is a loss of 0.028 a.m.u that is converted into energy. Also 2 positrons and 2 neutrinos are released in the reaction.

surfaces of neighbouring grains, allowed them to agglutinate into larger and larger clumps, much as ice cubes adhere to one another in ice buckets. When the central mass increased, so did the angular velocity of the disk and grain collisions became more frequent and impacts more energetic, thus increasing the grain temperature as they collided. This temperature increase started to vaporize part of the gases frozen on the grain, beginning with those with a lower boiling point, and progressively increasing the gas content of the disk. The larger and larger grain cumuli now began to attract one another gravitationally, forming progressively larger and more compact bodies, the planetesimals and protoplanets, that accreted essentially all the material present in their neighbouring zones, until a planet or a smaller asteroid was formed. It is thought the large gaseous planets, Jupiter and Saturn, formed first, while the smaller rocky planets, Mercury, Venus, Earth and Mars formed later. When the Sun started hydrogen ignition it generated a shock wave –the T Tauri wind– that blew out the remaining disk gas and dust grains from the solar system.

Figure 3: A PROTOPLANETARY DISK

The central protosun and the rotating protoplanetary disk formed by gas and cosmic grains is shown. (Credit: NASA/JPL-Caltech).

Nuclear Fusion in the Sun

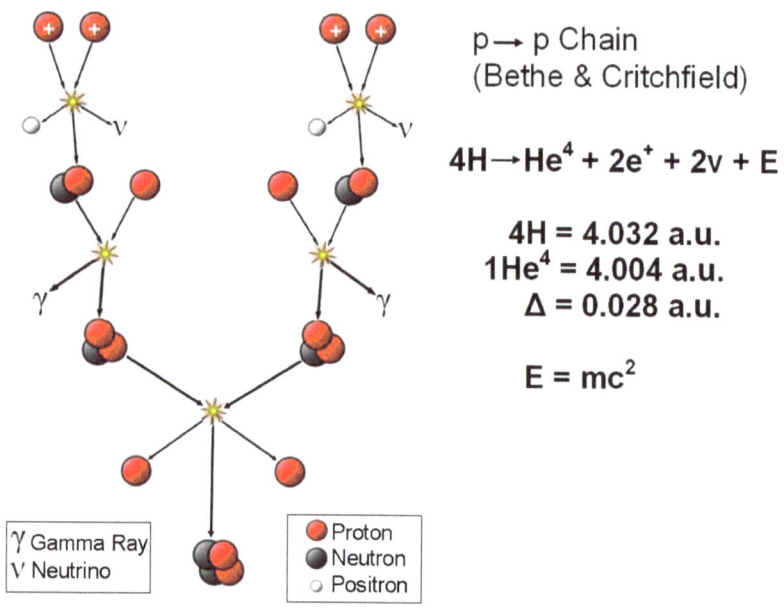

p → p Chain
(Bethe & Critchfield)

$4H \rightarrow He^4 + 2e^+ + 2v + E$

$4H = 4.032$ a.u.
$1He^4 = 4.004$ a.u.
$\Delta = 0.028$ a.u.

$E = mc^2$

γ Gamma Ray
ν Neutrino

● Proton
● Neutron
○ Positron

Figure 4: NUCLEAR FUSION IN THE SUN

This is the main energy-generating process in our Sun. H nuclei (protons) first fuse to generate heavy hydrogen (deuterium, ^2H) with the release of positrons (positively charged nuclear electrons), and neutrinos. Then, further fusions with protons leads to the formation of He of mass 4, with the release of protons and γ-rays, and the production of energy because of a decrease of mass by 0.028 atomic mass units (a.u.). (Credit: Wikimedia Commons).

Also, when the Sun started to burn hydrogen and release energy, this increased the outwardly directed pressure of its gas until it balanced the gravitational pull that tried to contract it. Thus an equilibrium was reached that shall last until the Sun exhausts most of its fuels in some 5000 Myrs when it shall expand to become a red giant.

The presence of a mass as large as the Sun at the center of a rotating disk, should, because of the conservation of angular momentum, lead to its faster rotation, that would rip the Sun and the entire disk apart, making the system unable to keep its structure and to evolve, eventually, till life appeared. This did not occur, however, because of the existence of a braking system in which angular momentum was transferred from the Sun to outer regions. Although this transfer mechanism is not well understood, it may involve a magnetohydrodynamic process and, perhaps, also the emission of matter in the form of molecular outflow jets, containing H, CO, CO_2 and other gases, that may extend for great distances, interacting with distant interstellar masses [12,15,16].

CHAPTER 2

A NEW EARTH AND A NEW MOON

BUILDING THE EARTH

The Earth is now known to have formed from the progressive accretion of smaller elements, starting with the submicroscopic interstellar dust grains that formed the protoplanetary disk. During the initial stage the dust grain clumps were still cold and able to keep in their interior the frozen gases that had a higher melting temperature, including water. As the accreted masses became larger and the disk rotated at higher speed, the collisions among them became more energetic and their temperatures increased. When planetesimals and protoplanets formed the high temperatures attained after the accretionary shocks were able to melt the newly formed bodies. In this molten state, the protoplanetary components differentiated according to their densities, with the denser ones, like Fe and Ni, sinking to the center of the protoplanet, and the lighter ones, such as Si, Al, Mg, C and others floated to more superficial regions. Once the differentiated body cooled, the new distribution of materials persisted. Also as the submicroscopic dust grains that accreted into the solid bodies of the solar system were heterogeneous, all the bodies formed from them were also heterogeneous[1,2].

In a further step, the large planetesimals present in a given zone also started to accrete, in a process that lasted some Myrs, giving rise to several molten planets, amongst them the first Earth[3,4]. This accretionary theory of the Earth's formation was first put forward by the Bielorussian scientist Otto Schmidt (1891 – 1956) and is now widely accepted[5].

DIFFERENTIATION OF THE EARTH

As mentioned, when the Earth was in a molten state the denser elements such as Ni and Fe and the "iron-loving" (siderophile) components formed the planet core, while the lighter ones floated to form the surface crust. The components with intermediate density segregated into the mantle, the semi–solid, plastic region lying between the core and the crust[3].

This initial differentiation was of paramount importance for the appearance of life. The presence of a solid inner core, made of iron, nickel and long–lived radioactive elements such us uranium of mass 238 (^{238}U), thorium (^{232}Th) and potassium (^{40}K), and surrounded by a large fluid outer core, that also contained iron, was responsible for the generation of the Earth's strong magnetic field, that, in turn, generated the magnetosphere, the shield that surrounds our planet, and deflects the solar wind, the 400 km/s flux of high energy protons and electrons generated in the Sun[7], that would have destroyed living beings on the Earth's surface, as well as blowing away the primitive atmosphere, as was the case with Mars, that lacks a significant magnetic field. Such solar wind could have been initially up to 1000 times stronger than at present[8].

On the other hand, the less dense components of the molten Earth, floated to the surface and gave rise to the rocky crust, another essential element for life, because the crust is impermeable to water, and the rocks that form it will evolve, by fragmentation and comminution, to form soils[9].

The components of intermediate density, mostly iron and magnesium silicates, the so called mafic and ultramafic (a name derived from magnesium and ferrum) minerals formed the semi-solid plastic mantle that transmits to the crust the heat generated in the core, mainly by means of convection[3]. This convective activity is what drives plate tectonics, the mechanism by which the emerged land travels on the planet's surface, and is kept above the ocean's level. If this did not exist, all land would be 2700 m. under the oceanic surface and only marine life would be possible.

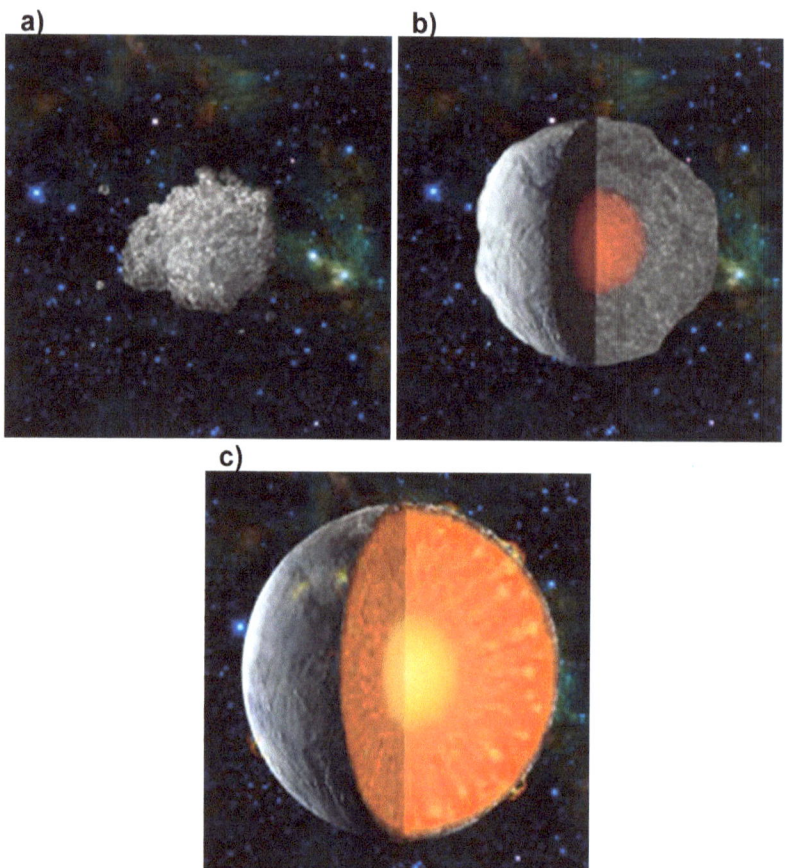

Figure 5: **PLANET DIFFERENTIATION**
(a). A clump of cosmic dust granules, formed mainly by core-mantle grains. These clumps grow and impact with other grain clumps until larger rocky masses form (b). These larger masses (planetesimals) heat and start to differentiate (c), with the heavier elements (Fe and Ni) sinking to form the core, while the lighter ones float giving rise to a crust, and those of intermediate density forming the mantle. (Credit: JPL/NASA).

Thus, it may be seen that the differentiated Earth forms a coordinated system whose different layers interact to give a functionally unified structure in which life shall eventually arise. Also the density-driven partition of materials represents a natural selective mechanism present in the mineral realm.

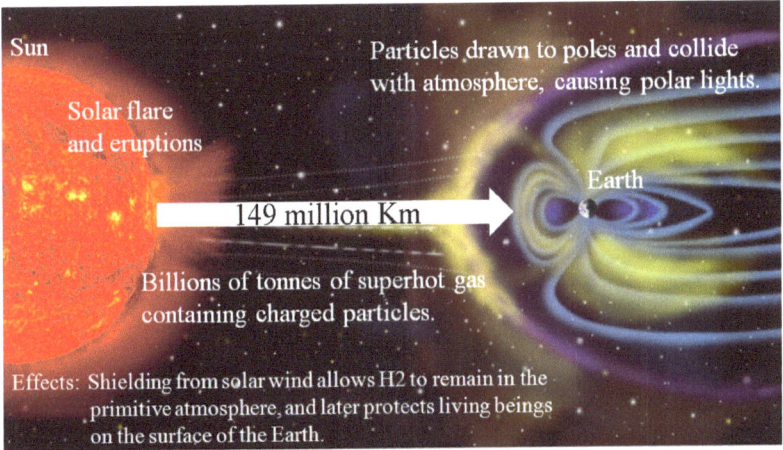

Figure 6: SOLAR WIND
The sun emits billions of tons of electrically charged particles (electrons, protons and helium nuclei) that travel at high speed throughout the Solar System. The Earth, because of the magnetic field generated by its iron nucleus, deflects most of this "wind". (Credit: NASA).

A NEW MOON

The Earth had finished its initial accretion to 80-90% of its present mass at about 100 Myrs following the beginning of its accretion process[3]. Then, at 4480 Myrs before present (b.p.), the most catastrophic event ever in Earth's history occurred: a Mars–sized planet crashed onto the Earth. This impact was off–centered and increased the speed of rotation of the Earth. The energy released was so great that both the Earth and the impactor (sometimes called Theia [note 3]) melted completely, and the impactor's differentiated iron core sank into the Earth, increasing the mass of its core, while its molten silicate-rich outer layers mixed with the Earth's molten or vaporized

Note [3]: From Greek mythology; Theia was said to be the mother of the Moon (Selene) as well as of the Sun (Helios) and of Eos (the Dawn).

fragments, and were ejected into space by the force of the impact. Part of these fragments settled back onto the Earth, while those that had reached a higher altitude, beyond what is known as the Roche limit, started to coalesce gravitationally, leading to the eventual formation of the Moon.

This "new" Moon stabilized the tilt of the Earth's axis, the orientation of which was previously chaotic, allowing seasons to occur. Also it increased the Earth's core, thus, strengthening its magnetic field. The increase of the planet's mass (to 99% of its present value) allowed the retention of a larger atmosphere, and also positioned the planet in a larger orbit, well into the habitable zone of the solar system, not too close to the Sun, where it would be too hot, nor too far, where it would be too cold, thus making life possible.

Figure 7: **THE ORIGIN OF THE MOON**

Upper Left: A Mars-sized planet (Theia) impacts the Earth. Middle: As a result of the impact both bodies heat and shattering of Theia and of part of the molten Earth occurs, with partial mixing of their contents. The shattered fragments as well as a silicate dust atmosphere circle around the new Earth, with those closer to it finally falling on it, while the more distant one crash among themselves, increasing their size until a molten Moon is formed. (Lower right), (Source: bbc.co.uk).

The Earth and Moon now constituted a dynamical system, in which angular momentum must be conserved. Initially, the rotation period of the Earth (the day) lasted 5 to 6 hours. As the Moon progressively receded, days started to get longer, being now 24 h, and still increasing slowly.

The newly formed Moon was iron-poor (only 10% of its mass was iron, against the Earth's 30%), had a small core, and 75 – 90% of its mass derived from the impactor, while 10 – 25% of it had a terrestrial origin. Its mass was only 1.2% of that of the Earth, so it could not retain a gaseous atmosphere, and its temperatures are presently extreme, with a maximum of 123°C and a minimum of -233°C.

PREPARING A CRADLE FOR LIFE

a. Temperature of the Earth

During the very initial moments of our planet's accretion, the circum-solar disk rotated slowly and the minute dust grains collided gently among themselves, without a significant change in temperature. As the grains formed larger clumps and the disc rotated more rapidly, the collisions became more energetic and the temperature increased progressively. When planetesimals and protoplanets formed, the energy of the collisions among them became high, and this induced their melting and the subsequent differentiation of its components. So, the proto-Earth became intermittently very hot, especially during the period when the accretion rate was high. As the planet approached 70% of its present mass, it had cooled by radiation, and was formed by a well differentiated core, mantle and crust. The great Moon-forming impact released a huge amount of energy that rewarmed the Earth, allowing part of its more superficial layers to shatter and mix with fragments of Theia, and part of the Earth's silicates to vaporize, forming a 4000°C silicate vapor atmosphere, while the surface became a magma ocean of molten rocks.

The appearance of a very hot magmatic Earth had two very important biotropic effects. One was that the superficial layers were molten, and when the planet cooled, in some 2 Myrs, the molten rocks started to crystallize progressively in a well defined order, known as the Bowen series, in which rocks containing calcium, magnesium and silicates crystallize first, i.e. at higher temperature, followed by Ca and Na rich plagioclase feldspars, and, at lower temperatures by Na–rich plagioclase feldspar and by K, Mg, Fe, Al and Si-rich biotites. Still at lower temperatures, K, Al and Si – containing minerals (potassium feldspars and muscovite) crystallize, while quartz (SiO_2) is the last to crystallize.

The second biotropic consequence was the generation of a second atmosphere.

b. Earth's Atmosphere

The first atmosphere of the protoEarth was composed mostly of hydrogen and, in a lesser proportion, by helium and other minor gases that, because of their very low freezing temperatures, had not been incorporated into the frozen mantle of the interstellar dust grains. This first atmosphere was made of very light elements held by the Earth's gravitational field, and protected from the solar wind by its magnetic field. Following the Moon-forming impact, however, the high temperature generated (4000°C or more) and the turbulence produced by swirling rock fragments, blew off this first atmosphere. Next, the molten Earth, without its insulating crust, like a gigantic volcano, was able to degass many of the gases trapped in it since its origin. Thus, a second atmosphere was generated, formed by rocky dust (that lasted 1000 yrs.), and by hot water vapor (steam), CO_2, carbon monoxide (CO), methane (CH_4), nitrogen (N_2), sulfhydric acid (H_2S), carbon oxysulfide (COS), ammonia (NH_3) and other minor components, but no hydrogen or oxygen. All these gases had been present in the mantle of the original dust grains. The atmospheric pressure on the Earth's surface was very high, perhaps 200 bars or more, indicating a high density and a volume many times larger than the present one.

When the crust started to reform, outgassing continued due to intense volcanism, that progressively decreased with diminishing temperature and with the growth of the crust.

CHAPTER 3

ON LIGHTNINGS AND BOMBS

STANLEY MILLER'S GREAT EXPERIMENT

In 1951, Stanley Miller (1930-2007) a young graduate student at the Department of Chemistry of the University of Chicago, attended a lecture by Harold Urey, the discoverer of deuterium (^{2}H) and a Nobel Prize winner, on the atmosphere of the planets, including the Earth, which was thought to have been initially composed of hydrogen, methane and ammonia. After the lecture, Miller approached Urey and suggested performing experiments in which spark discharges were to be applied to a mix of H_2, CH_4, NH_3, and H_2O, hoping that life-related substances would be formed. Urey agreed reluctantly, giving Miller a 6 – 12 month period to perform it. After a few weeks of experimental work, Miller found that amino acids, especially glycine, α -and β- alanine and perhaps also aspartic acid and γ-aminobutyric acid had been formed after one week of spiking the gas mix with spark discharges generated by a Tesla coil[1,2]. Further analysis showed that other amino acids, as well as aliphatic and hydroxylated acids, urea and other undefined compounds had been formed in a surprisingly high yield: In all, 15% of the CH_4 carbon had been transformed into organic molecules that now form part of living organisms, with glycine accounting for 2.1% of CH_4 carbon. Reanalyzing Miller's original samples with modern analytical methods has recently shown that 22 amino acids, as racemic (DL) mixtures, five amines and several hydroxylated compounds had also been formed[3]. (Table 3-1)

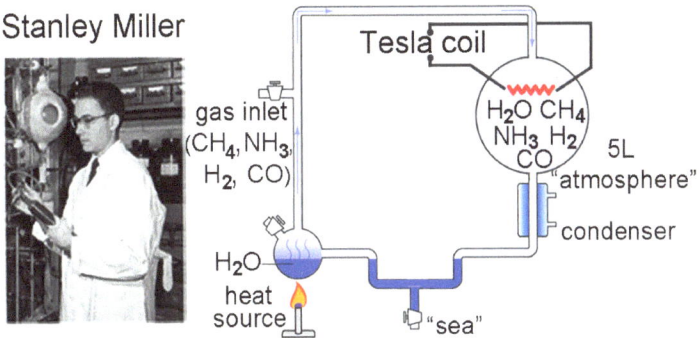

Figure 8: STANLEY MILLER

The photograph of S. Miller, when he performed his famous experiment, is shown next to the apparatus he used.

These experiments caused a great impact and were quickly replicated using different gas mixes, provided that they were of a reducing character, as the presence of free oxygen precluded all synthesis. This need for a primitive reducing atmosphere was, however, a large stumbling-block for Miller and Urey's theory, because most geophysicists believed, until recently, that the atmosphere was then a CO_2-rich, oxidizing one. In recent years, however, this has changed entirely because of experiments made with the degassing of different types of chondrites (meteorites), of which "ordinary" chondrites constitute more than 63% of all chondrites. These are fragments of small asteroids formed by the accretion of dust grains present in the pre-solar nebula, that contain metallic (i.e. reduced) Fe and Ni in variable proportions, depending on the chondrite subtype, as well as silicates and other minerals, and are also rich in volatiles. When ordinary chondrites were heated up to 1225°C, they released CH_4, H_2, H_2O, N_2 and NH_3; and when heated to higher temperatures, CO replaced CH_4 in the gases released. That is, in all cases the gases released were of a reducing character[4]. Something similar was seen with the less abundant (2% of total), enstatite chondrites and with CV carbonaceous chondrites, while most carbonaceous chondrites, that represent only 2.5% of all chondrites, release CO_2 on heating. All three similar studies performed independently reached the same conclusion: "If the gases of Earth's earliest" (i.e. secondary) "atmosphere had equilibrated with material like that in primitive meteorites, the atmosphere that results would be much more reduced than modern volcanic gas"[5,6].

TABLE 3-1
ORGANIC COMPOUNDS FORMED IN SPARK DISCHARGE EXPERIMENTS

AMINO ACIDS	ACIDS AND AMINES
Glycine	Urea
Alanine	Formic Acid
β- Alanine	Acetic Acid
Serine	Propanoic Acid
Isoserine	Glycolic Acid
Homocysteic Acid	Lactic Acid
β- Aminobutyric Acid	Succinic Acid
α- Aminobutyric Acid	α- Hydroxybutyric Acid
α- Aminoisobutyric Acid	Methylamine
γ- Aminobutyric Acid	Ethylamine
β- Aminoisobutyric Acid	Cysteamine
Threonine	Ethanolamine
Aspartic Acid	
S- Methylcysteine	
Valine	
Isovaline	
Glutamic Acid	
Methionine	
Methioninesulfoxide	
Methioninesulfone	
Proline	
Isoleucine	
Norleucine	
Leucine	
Ethionine	

As we have seen, our planet was formed from those same small dust granules that formed chondrites, and that such granules were present in the primordial solar nebula, where they were surrounded by reducing hydrogen gas. Also there was plenty of metallic iron and nickel in the grain cores, and the majority of their frozen mantles contained numerous reduced molecules (e.g. CH_4, CO, H_2O, H_2S, NH_3, CH_3OH, H_2CO, C_2H_2 and others). It is, thus, readily understandable that the Earth contained, especially in the deepest regions of the crust and in the mantle, a majority of reduced substances, albeit many of them could be in a latent state, e.g. water adsorbed to minerals, or as water of hydration, and gases simply kept in bubbles trapped in rocks.

As with other fluids, gases tend to distribute themselves according to density. Thus CO_2, the heaviest gas (m.w.=44; density 1.977 kg/m^3 at $0°C$ and 1 atm pressure) would be at the bottom, and H_2, the less dense one (0.0899 kg/m^3), at the atmosphere's top, from where it could escape easily. In progressively deeper atmospheric layers, from top to bottom, we find CH_4 (0.717 kg/m^3), NH_3 (0.769 kg/m^3), water vapor (0.804 kg/m^3); CO and N_2 (1.25 and 1.2506 kg/m^3, respectively), while H_2S (1.434 kg/m^3) and CO_2, as mentioned, lie at the lower levels. Although these relative positions, due to turbulences, atmospheric currents and thermal effects, are not fixed, they represent regions where finding a given gas is most probable.

It has been speculated that water on Earth was mainly brought by comets, that are rich in it. However, the deuterium to hydrogen ratios (D/H) are almost identical between the oceans and water contained in the Earth's mantle, while the D/H ratio in comet water is only one half of the former, with the exception of CI carbonaceous chondrites, that represent a very small fraction of all chondrites, and of comet 103P/Hartley 2, where the D/H ratio is similar to that of the Earth's oceans. Moreover, the D/H ratio of water on Earth and on the Moon's surface and interior is the same, thus favoring an early accretionary origin of water on the proto-Earth before the occurrence of the Moon-forming giant impact. This seems to favor an ancient accretionary origin, as the one postulated here, for the origin of our planet's water[7, 7a, 7b].

BOMBS FOR LIFE: THE LATE HEAVY BOMBARDMENT (LHB)

In this hot barren world, the emergence of life was not possible. But then, one of the most extraordinary events in the history of our planet began: the Earth, the Moon and the rocky planets in the inner solar system as well as part of the outer planets started to receive an intense shower of asteroids and comets originating mostly in the main asteroid belt. This intense bombardment is thought to have lasted from about 4100 to 3850 Myrs b.p., and to have been triggered by changes in the orbits of the giant planets Jupiter, Saturn, Uranus and Neptune, so that finally, Neptune shifted with Uranus their orbital trajectories, and Neptune, now the outermost planet, penetrated the planetesimal disk, dispersing gravitationally 97% of its mass, and placing its asteroids in highly elliptical orbits that reached into the inner solar system, giving rise to its prolonged bombardment[8,9].

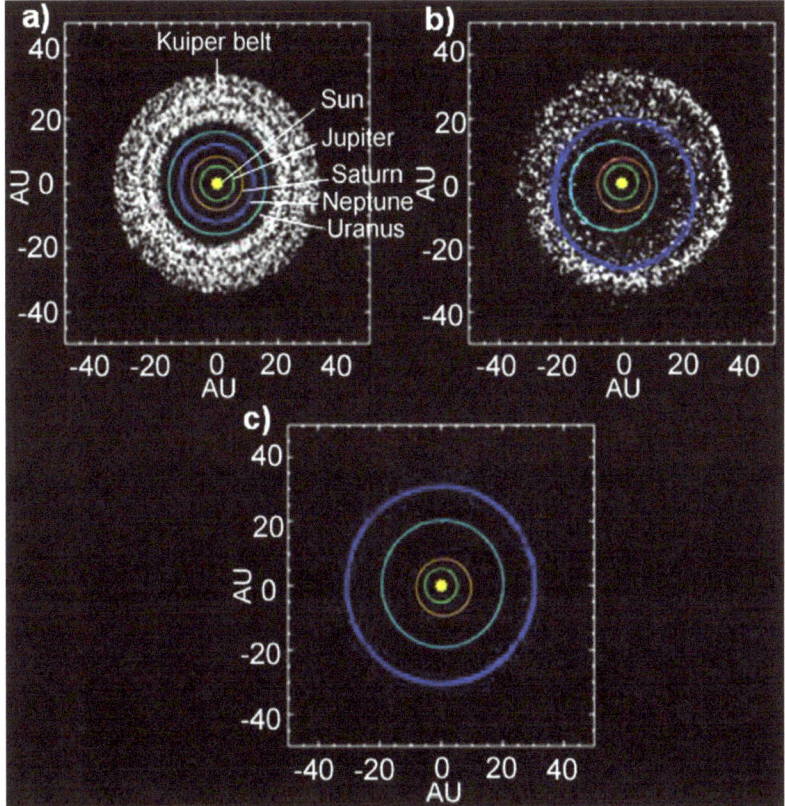

Figure 9: GASEOUS PLANET MIGRATION AND THE GENERATION OF THE LATE HEAVY BOMBARDMENT

The positions of the Sun, of the large gaseous planets and of the Kuiper asteroid belt are indicated. Due to an interaction between the orbits of Saturn and Neptune, that of the latter migrates outwardly, penetrating the asteroid belt and throwing most of its asteroids into elliptical orbits, leading to the bombardment of both the rocky inner planets and of the gaseous ones as well as their satellites. AU is one astronomical unit i.e. the distance from the Sun to the Earth. a) Early configuration of outer planetary orbits. b) Scattering of asteroids into the inner solar system following the shift in the orbits of Uranus and Neptune. c) After the ejection of the asteroids. (Credit: Wikimedia Commons, modified).

Consequences of the LHB

a. On the Atmosphere

When an asteroid enters the Earth's atmosphere, heat is generated by the compression it induces in the air in front of it (the ram pressure), which may reach 4100°C. At this temperature a plasma is formed in the air and on the meteor surface. This temperature decreases with radial distance from the meteor. It is known that the temperature reached is largely independent of the meteor's mass, while the volume of air heated is a function of such mass[11]. Thus, during the LHB, that lasted several hundred million years, an extremely large total mass of extraterrestrial matter penetrated the atmosphere, heating large volumes of it.

Heating of a methane-rich steam atmosphere could lead to hydrogen generation by a process similar to the industrial steam reforming process[12]. In it, heating methane with steam at 700-1100°C, yields hydrogen with 80% efficiency ($CH_4 + H_2O$ (steam) $\longrightarrow CO + 3H_2$).

At lower temperature (130°C), CO reacts with water to form $CO_2 + H_2$. Such processes, as well as direct electrolysis of water at 4100°C and others not yet well defined, would have produced a hydrogen-rich reducing upper atmosphere that lasted as long as the LHB did. In addition, the intense lightning activity present at the time, that generates temperatures of 20,000°C or more, would also generate large amounts of hydrogen in the methane–water vapor atmosphere[5-10]. Thus, it may be seen that there was plenty of reducing power in the early atmosphere and that the criticisms to the Miller-Urey experiments were entirely unfounded.

b. On Land

No traces of the direct actions of LHB on the Earth's crust remain. This is because of the erosion of the original rocks and, especially, because of subduction of the crust into the Earth's mantle due to plate tectonics.

In the Moon, however, because of the absence of winds and rain, there has been no significant erosion; and also, the absence of major plate tectonics has allowed the crust present during the LHB to remain largely unmodified. Thus much of the action of LHB such as cratering, rock mixing, melting and overturning, as well as their fragmentation and the production of rubble (regolith), remain visible, together with the effects of some volcanism.

The large number of craters seen on the Moon's surface, ranging from very large ones, such as the South Pole-Aitken basin, 2600 km in diameter, to some that measure a few cm., bear witness to the intense bombardment the Moon has suffered throughout its history, especially during the LHB. From distant observations, especially the Clementine, Lunar Prospector and Selene missions, as well as the lunar soil and rocks brought to Earth by the Apollo and Luna missions, and the study of lunar meteorites on Earth, a fairly good knowledge of the Moon's structure and history has been obtained, which, in turn, has clarified fundamental aspects of the Earth's history[13-16].

It is now generally accepted that, following the giant Moon-forming impact, the Moon and a large part of the Earth became a hot magma of molten rock: the Magma Ocean. And, as heat was lost to space, the temperatures of both bodies decreased, and the rocks started to crystallize, following the Bowen crystallization sequence. This process led to magmatic differentiation as the less dense calcium plagioclase anorthite ($CaAl_2Si_2O_8$) floated to the surface forming a crust, as has been found by Ohtake et al[16]. In turn, mafic minerals, due to their higher density sank to the bottom, forming a mantle.

Figure 10: **A Battered Moon**
The back side of the Moon showing the large number of depressions produced by the impact of asteroids and comets during its long history, especially during the Late Heavy Bombardment. Because of the absence of winds and rain these impacts have not been erased. (Credit: NASA/ GSFC/Arizona State University).

The finding by Ohtake et al., using the Japanese SELENE explorer, that essentially all of the lunar highland surface, that represents the most ancient and unmodified lunar crust, is formed by the calcium plagioclase anorthite, is of great importance for the reconstruction of the Earth's early, post–Theia impact, as the main geological features of Earth and

Moon were essentially similar at the time. This implies that the Earth's crust in this period was also formed by this mineral, although there are no vestiges of it.

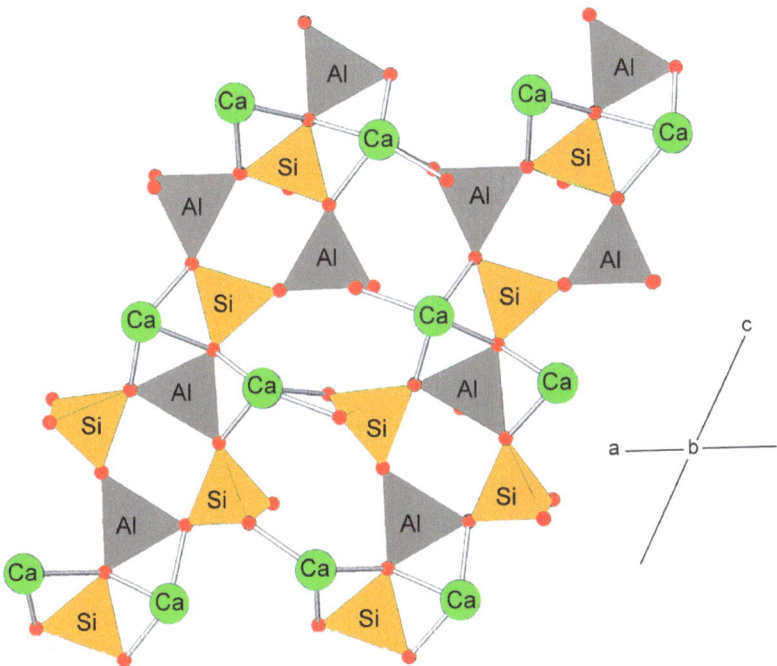

Figure 11: ANORTHITE STRUCTURE
Calcium (green), silicon (yellow), aluminum (gray) and oxygen (red) atoms are shown. (Credit: M. Sandoval).

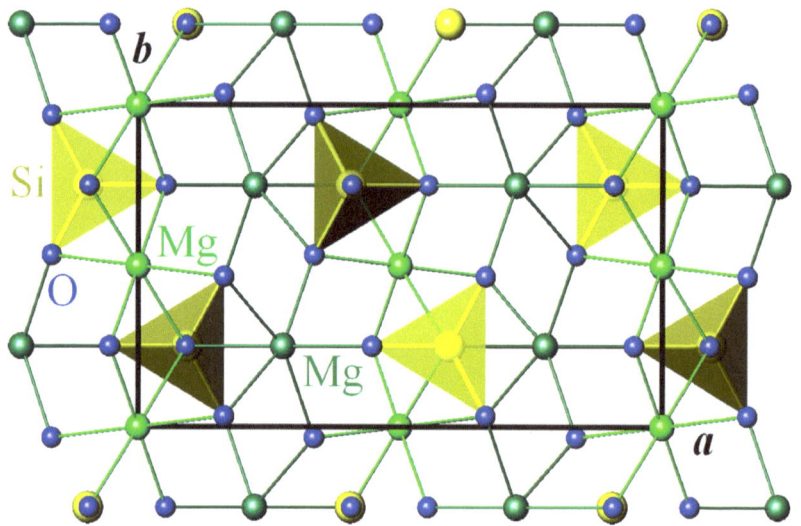

Figure 12: OLIVINE CRYSTAL STRUCTURE
Mg (green), Si (yellow) and oxygen (blue) atoms are shown. (Credit: Wikimedia Commons).

There is still a third group of minerals formed by the so called "incompatible" elements such as potassium (K) and thorium (Th). These elements fit with difficulty in crystal structures and thus remain segregated between the crust and the mantle, almost certainly in discrete magma chamber reservoirs[15]. K is associated in them not only with Th, but also with rare Earth elements (REE), such as gadolinium and samarium and other fifteen elements of this group, and with phosphorus (P) oxides. This has led to the acronymic naming of KREEP for these minerals. It has also been shown that uranium (U) is present in KREEP[15]. As K, Th and U are long-lived radioactive elements that heat their surroundings (they are currently the main heat source for the Earth's interior), it is possible that they were responsible for keeping the KREEP reservoirs in a molten state.

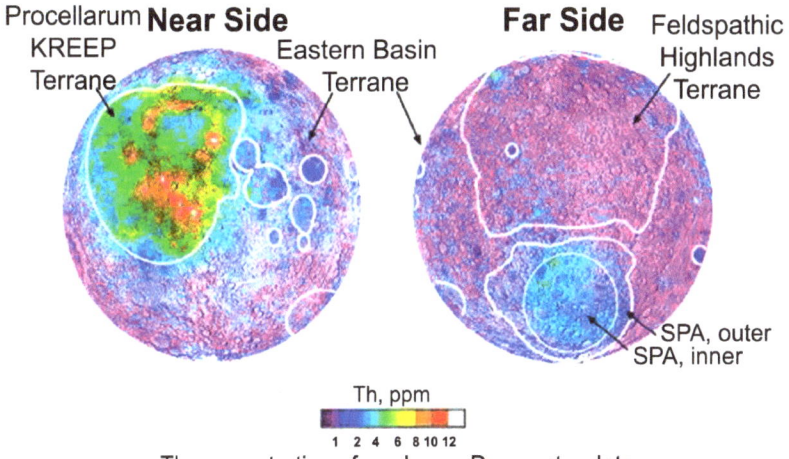

Th concentrations from Lunar Prospector data,
calibrated to landing site soils (Gillis et al., 2000).

Figure 13: CHEMICAL COMPOSITION OF THE MOON

The different terranes on the Moon crust are depicted. The KREEP Terrane is rich in potassium (K), phosphoric oxides (P) and thorium (Th). Most of the surface is covered with anorthosite (anorthite), a feldspathic mineral. Data obtained by the Lunar Prospector and Clementine missions. (Courtesy of J. Wiley and Sons).

c. Crater Formation on Earth

The Earth, 80 times more massive than the Moon, should have attracted LHB impactors of much larger size and frequency than the Moon. It has been estimated that a crater as large as a continent, as well as many others of smaller size, were generated during the LHB. These craters led to excavations proportional (up to 10%) to their diameters[17] that could have reached the upper mantle and the KREEP chambers in it, with the subsequent extrusion of these minerals. This happened in the Moon as shown by the presence of vast KREEP terrains, in which the original crust was replaced by these minerals, and also by the examination of the lunar rocks brought to Earth by Apollo astronauts and, especially, of the lunar meteorite SaU 169[18,19].

Another of the effects of the impactors was to heat the impact area. It is known that heating phosphates, as those in KREEP, to 700-1000°C, leads to the formation of polyphosphates, with up to several thousand orthophosphate monomers, linked by high-energy phosphoanhydride bonds, in which physical energy (momentum and heat) is transformed into high-energy chemical bonds[20,21]. Polyphosphate was postulated by Jones and Lipmann to be the primeval source of energy for early forms of life[22]. With our present knowledge I think this postulate emerges now as virtually certain. We may add that, as we shall show here, it was also the main source of chemical energy for prebiotic synthesis.

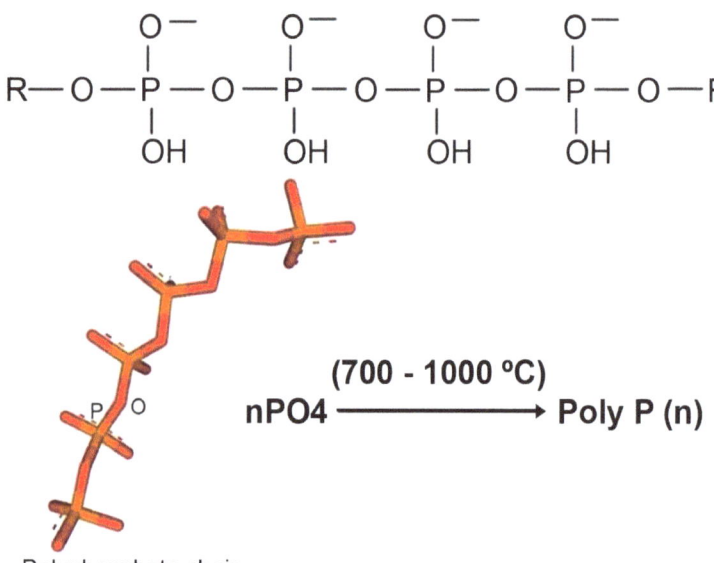

Polyphosphate chain
-377.114 kcal/mol

Figure 14: **POLYPHOSPHATE**
The structure of this polymer, formed by orthophosphate subunits linked by acid anhydride (high energy) bonds is shown. In it thermal energy is converted into chemical potential energy, capable of driving many chemical endergonic reactions, i.e. that need the supply of exogenous energy to occur. (Credit: Carlos Lagos, Centro de Bioinformática, Univ. Católica de Chile).

CHAPTER 4

LIFE PRECURSORS FALL INTO PLACE

WATER'S MANY BIOTROPIC EFFECTS

When the temperature of the Earth's atmosphere decreased following the Moon-forming impact, the atmospheric steam started to condense into water drops and intense rains followed. But, as the land was still hot, this water re-evaporated back into the atmosphere. On evaporation, water molecules gain thermal energy that cools the surface where evaporation takes place. As water molecules ascend in the atmosphere, that becomes progressively colder with altitude, a point shall be reached (the dew point) where they condense again releasing the thermal energy absorbed during evaporation. This released energy shall "energize" the clouds and induce, by a mechanism that is not clear, lightning production. The generation of lightning is today much more intense in tropical regions, where water evaporation is greater, thus indicating its connection with water vapor condensation. In the early period of a hot atmosphere, with high water evaporation rates, the intensity of lightning must have been truly hellish (Hadean) [note 4].

The biotropic effects of lightning and thus of water, are multiple. The temperature of the air close to a lightning bolt may reach 20,000°C. This, as mentioned, in a methane-water vapor- and CO- rich atmosphere shall generate large amounts of hydrogen that, together with the hydrogen generated by meteors during the LHB, suggest that there was no H deficit and that the Miller-Urey experiment did, in fact, occur in Nature in a vast scale, especially if one considers that in the original

---

Note [4]: Hades, the Greek mythological god of the underworld, part of which (the Phlegethon river) was a place of fire. Because of its fiery environment the early Earth has been called the Hadean era.

Miller experiment the "atmosphere" (i.e. the flask) had a 5 liter volume, against some 3×10^9 km^3 of the real atmosphere at the time, and that Miller's experiment lasted 7 days, against some 200 Myrs of the natural experiment, not to mention the differences in the energy applied in both cases.

THE FALL OF CO$_2$

The calcium and aluminum silicate anorthite, that covered the Earth's crust, is quite susceptible to weathering, especially by moderately acidic hot water. During the rains of the late Hadean, hot water drops dissolved the abundant atmospheric CO_2, that in solution forms carbonic acid (H_2CO_3), that is in equilibrium with H^+ and HCO_3^-. This hot acidic rain weathers anorthite ($CaAl_2Si_2O_8$), with the formation of calcium bicarbonate, that, when water becomes liquid, will be washed away into the water of lakes and seas. But Al_2O_3 also reacts with water forming $Al(OH)_3$, an alkaline substance that is also dissolved and washed away. This will make the waters alkaline and calcium bicarbonate ($Ca(HCO_3)_2$), that is relatively soluble, will become $CaCO_3$, a highly insoluble substance that precipitates to the bottom[1,2]. This mechanism, because of the abundance of warm rain and the ease with which anorthite is weathered[3], should rapidly reduce the concentration of CO_2 in the atmosphere, making it more reducing and thus still more favorable for the synthesis of organic molecules.

The study of Hadean zircons older than 4000 Myrs has shown that the temperature in the lower crust, where these small crystals were presumably formed, was 650-800°C, and that in the upper mantle the temperatures were higher[4]. As we have seen, polyphosphate is formed at these temperatures. Also in the chambers where KREEP was present, the radioactivity of K, uranium and thorium, also present in KREEP, would further heat the phosphates increasing the rate of polyphosphate formation. We can, therefore, predict that polyphosphate will be found in the lunar KREEP terranes and in the KREEP rocks brought to Earth in the Apollo missions.

SYNTHESIS OF ORGANIC MOLECULES IN THE HADEAN ATMOSPHERE

Lightnings generate electromagnetic radiation that go from infrared to visible, UV, X-rays and energetic gamma-rays, that are even able to give rise to matter-antimatter particles[5]. This radiation is the equivalent, but to a far greater scale, to the spark discharges of the Miller experiment, that caused atomic reshuffling and, thus, the synthesis of organic molecules.

As many molecules such as aminoacids, metabolic intermediates, fatty acids, nucleic acid bases and others, were synthesized in the Miller – Urey and other abiotic experiments[6,7], it may be reasonably postulated that most of the essential monomeric molecules were formed in the atmosphere during the millions of years in which it was energized and had a reducing character. These molecules were eventually washed by rain onto the Earth's surface where they remained with little degradation, due to the absence of living beings, as Darwin recognized[8].

In the original Miller experiment, during the seven days it lasted, about 15% of the methane present was converted into organic matter. In nature, where there was a high concentration of atmospheric methane, it may be safely assumed that the amount of organic molecules formed was indeed very large. These synthetic mechanisms, together with the H-generating one, were responsible for the decrease of methane in the atmosphere.

THE SEA WHERE LIFE WAS BORN

Early in the 20th century biochemists, especially McCallum, had recognized that the extracellular fluid of vertebrates, including Homo, was similar, in its inorganic ion composition, to the ocean where those vertebrates first appeared[9,10]. Following the same line of reasoning, it seems logical to think that the ionic composition of the simplest uni-cellular organism, the prokaryotes, must have been similar to the sea where these organisms first appeared. And as this intracellular fluid is rich in potassium (about 150 mM) and phosphate, with minor amounts of magnesium and iron, we can postulate that the sea in which the first living beings appeared was also rich in K and phosphate (mostly polyphosphate), with minor Mg and Fe components. And, in the same manner as our extracellular fluid ("the sea within us" as Epstein called it[11]) has remained constant for millions of years, our intracellular fluid is a relic of that present in the first cells, born more than 3600 Myrs ago, and of the ocean that bathed them. It is "our intracellular ocean". Also, McCallum reported that the Archean sea had a K concentration that was up to 2.5-fold higher than that of sodium[10]. This suggests that our "biogenic" sea was even richer in K.

It is also appropriate to consider the Darwinian principle of com-mon descent, by which the presence of a given homologous character (in this case the intracellular ion composition), in a variety of different species, indicates that this character was present in a common ancestor (in this case, the first living cell).

The water that poured onto the Earth during the late Hadean, filled most of the depressions formed during the LHB, thus creating ponds, lakes, seas and oceans. In zones where KREEP was present in appropri-ate amounts (as its concentration was not uniform over the Earth's sur-face, similarly to what is observed in the Moon, where only a fraction of the surface is KREEP-rich), water would weather (dissolve) the KREEP rocks, producing a K and polyphosphate solution in the biotropic con-centrations already mentioned. Also, due to the excavation produced by the LHB impactors, mantle rocks, especially the magnesium -and iron-rich olivine, became accessible to water and, as the latter is one of the rocks most susceptible to weathering[3], magnesium (Mg^2) and ferrous (Fe^{2+}) ions were also dissolved in the biotropic sea where the latter is thought to have given a greenish-blue colour to it. This K, Mg, Fe and phosphate-rich fluid was, and still is, important for the structure of water and for protein function, that are very different in sodium, calcium and chloride-containing fluids, such as the extracellular medium[12].

The temperature of this ancient sea has been estimated, by three independent methods, to have been about 75 to 80°C at 3500 Myrs and 55-85°C at 3500 to 3200 Myrs ago, with higher temperatures at earlier ages[13-15].

As to the size of this primordial sea, it could have been not as small as a lake, because the depth excavated by the impactor at that time would not have gone deep enough for the deep crust and superficial mantle to be exposed to the weathering action of water, nor as large as an ocean, in which the newly formed molecules could have become much too diluted. In conclusion, Darwin's "little pond" could have been of the size of a sea, perhaps like the present Black Sea.

CHAPTER 5
COMPLEX MOLECULES START TO FORM

During the atmospheric phase of molecule formation, we can postulate that most of the monomeric compounds now present on Earth were formed in large amounts. This depended on the time during which their synthesis took place, that is, several hundred million years, as well as on the availability of methane, nitrogen, carbon monoxide, H_2S, H_2O, and hydrogen, that was also very large, as well as on the supply of energetic electromagnetic radiation, generated by lightning, that was also intense and prolonged.

This over-abundant molecular production was washed onto the Earth's surface once the atmospheric temperature decreased sufficiently for rain drops to form. This molecular rain was also of great magnitude and fell indiscriminately on both land and sea. In the latter it may have formed a rather thick broth, more concentrated than the "hot, dilute soup" postulated by Haldane[1].

The hot sea where the first cells were to appear was agitated vigorously by wind caused by the fast rotation of the Earth. The day, following the Moon-forming impact, lasted 5 or 6 hours, and at 3800 Myrs b.p., about 15 hrs. The tides in open oceans were also larger than at present, due to the lesser distance to the Moon. However, in a sea of the size of the Black Sea, tides probably did not exceed a height of 30 cm. This agitation increased the interaction of different molecules and the high temperatures, in turn, also increased the rate of chemical reactions. Tides, although not great, may have also allowed for chemical reactions to occur on solid surfaces, where organic molecules were deposited during low tide, and were exposed to UV radiation that is known to favor both synthetic and destructive chemical reactions[2].

LIPIDS FOR LIFE

Lipids are one of the basic constituents of living beings. This is because most of their molecular surfaces are water repellent (hydrophobic), making them suitable to act as barriers to water and water-soluble substances. This allowed the existence of aqueous small-volume compartments in which the biochemical reactions that characterize living beings could take place. Otherwise, life process would be diluted by the enormous external volumes in which these small- volume units existed.

The basic units of lipids are the fatty acids. These are easily formed under prebiotic conditions. Thus, in Miller experiments, straight-chain organic "fatty" acids, with chain lengths of up to 12 carbons are formed. Also in several carbonaceous chondrites, linear monocarboxylic acids also of up to 12 carbons have been detected, as well as a host of dicarboxylic, branched-chain and aromatic carboxylic acids, several of them in substantial amounts[3-5]. These compounds are thought to have been formed during aqueous alteration of the parent asteroid from which the chondrite (meteorite) derived. Carboxylic acids of up to 35 carbons can also be formed from CO and H_2, by Fischer-Tropsch synthesis, or by its variant, the Kölbel-Engelhard process at the high temperatures that could have occurred in prebiotic hydrothermal vents[6,6a,7] in the floor of the biogenic sea.

Glycerol, also a key component of lipids, has been shown to be easily formed under prebiotic conditions, and its presence in the Murchison chondrite at a concentration of 160 nmol/g indicates the ease of its abiotic synthesis[4]. Choline, which forms part of present day phospholipids, could be formed in the reaction of ethylene oxide and thrimethylamine ($N(CH_3)_3$), in an aqueous alkaline medium, while ethanolamine, another phospholipid constituent, could also have been formed from ethylene oxide and ammonia. Also, its production in the Miller H_2S-rich spark discharge experiment was substantial[8].

a. Formation of More Complex Lipids

Formation of mono glycerol esters and of phospholipids has been shown to occur by direct dehydration under dry, hot conditions, sustained during weeks. This has been postulated to have happened in small sea water tide pools that contained glycerol, phosphate and cyanamide and covered by a fatty acid monolayer that, when dried by evaporation over a sand or clay bottom, would lead to their dehydrative synthesis[9]. But, as mentioned, because in the medium- sized biogenic sea where life possibly sparked, tides were small and because of the faster rotation of the Earth at that time, when days that lasted 15 hr. or less, tides occurred more frequently, and the existence of tide pools could not have been an important feature. In addition, because of the copious rains, dryness does not seem to have been of common occurrence. In fact, the warm, humid atmosphere at the time has been compared to a sauna.

b. The Amount of Lipids in the Biogenic Sea: An Estimate

If it is conservatively assumed that one gram of straight chain fatty acids (cross-section 0.2 nm^2) fell over one m^2 of the Earth's surface and, thus, on the biogenic sea, during one year, and as 100 Myr the duration of the prebiotic period during which the atmospheric steam condensed and washed organic molecules onto the surface, it may be shown that the surface occupied by these fatty acids was 60,000 km^2, greatly exceeding the area of one square meter on which the fatty acids were deposited. Although the actual amount deposited might have been far less than the one estimated here, it nevertheless seems clear that the surface of one m^2 could not accommodate as a monolayer this amount of lipids, and that it should have collapsed with the formation of subsurface micelles, of bubbles above the surface, and of a variety of vesicular structures, product of the self-assembly properties of different lipid species[12,13].

Lipid Chaos

It is also worth considering that, with the large amount of lipids deposited during the long prebiotic period, many combinations of lipid structures were formed, giving rise to a chaotic system, subject to natural selection. Thus fatty acid bubbles, that gave the biogenic sea a foamy appearance, were short lived, while more complex lipids, especially tetraether lipids, similar to those actually present in the thermoacidophilic archeon Sulfolobus acidocaldarius[14], that remain stable at 80°C, and choline-containing phospholipids with long chain fatty acids, in which the positively charged nitrogen of a choline makes a strong ionic bond with the negatively charged phosphate of a neighbouring phosphatydyl choline, could have been much more stable, and thus allowing them to be positively selected.

A New Protein World

As already mentioned, in the Miller–Urey experiments a high proportion of atmospheric methane was converted into organic compounds: 22 aminoacids were also formed with glycine accounting for 2.1% of the methane carbon consumed in one week. Extrapolating this from the 5 l flask of Miller's experiment to the large volume of the primitive atmosphere, and from the one to five week duration of Miller type experiments, to one of 200 Myrs, or more, in nature, it may be seen that the atmospheric accumulation of amino acids, in spite of physical destructive mechanisms (radiations, etc), must have been enormous.

With the collapse of the steam atmosphere, as was the case with many other organic compounds present in it, copious amounts of amino acids rained on the surface of Earth, including its biogenic sea. There polyP could have reacted with them, as mentioned, by forming the highly reactive aminoacylphosphates that were able to form peptide bonds with the α-amino groups of other aminoacids, and giving a growing point for randomly formed polypeptides of variable lengths and composition. In the experiments of Sidney Fox and his group, in which amino acid polymerization (i.e. proteinoids) was attained by heating them at 150°C or more, under dry conditions, a certain degree of enzymatic activity was found[15,16]. It is conceivable that under the mild aqueous conditions we postulate, far greater functional activities (i.e. enzymatic, structural or membrane-associated) could occur in the randomly assembled proteins. Once living cells appeared, with the generation of new enzymes, protein degrading enzymatic activity necessarily appeared, and the early randomly generated proteins, that had contributed so much to the generation of prebiotic structures, were degraded, leaving no signs of their existence, as ripples in a pond.

LIPID–PROTEIN COEVOLUTION

Of the many lipid structures randomly formed in the biogenic sea, some, as those formed from single-chain fatty acids, in which vesicular structures are formed when the dissociation constant (the pK_a) of the lipid monomer carboxylate roughly coincides with the pH of the medium. This could have been important in the formation of the first protocellular structures with fatty acids chains of 8 or more carbons[17,18].

Because of the great variety of lipid membranes present in extant living organisms, it is not possible to derive a common ancestor structure for those present in the first living cell. It may only be said that its structure had to be able to withstand the high temperature present at the time (e.g. 70 – 80 °C). As indicated above, one could think of tetraether lipids or of choline-containing P-lipids with two long-chain fatty acids, such as a behenic acid, attached to their glycerol, that would be able to resist the temperature of the environment.

Also, as lipids have a low density, they would float in an aqueous medium such as the biogenic sea, forming lipid bubble towers, unable to interact with water soluble molecules. This difficulty could be solved by the association of the primitive lipid membranes with the early randomly formed proteins. These primitive lipoprotein membranes, because of their higher density, could then submerge and allow significant interaction with other water-soluble molecules. Thus, under the prevailing conditions prior to life emergence, lipoprotein membranes were a necessity. In fact, they were the first cell structure to be formed and the basic element on which all future life features were built.

SUGAR FORMATION: THE RUSSIAN CONNECTION

The basic principles of carbohydrate abiotic formation stem from the work of the Russian chemist Alexander Butlerow (1828-1886) who first described the formation of sugars from formaldehyde ("dioxymé-thylène")[note 5], in what is now called, from the contraction of formaldehyde and aldose, the formose reaction, in which formaldehyde, in alkaline media, polymerizes to give glycolaldehyde and other sugars, including ribose, one of the components of RNA[19,20].

These experiments were later followed, in the early 20th century, by those of a number of English chemists that were able to show that UV light could convert CO_2 in aqueous solution, in the presence of colloidal $FeCl_3$ or colloidal silicic acid, into formaldehyde and also into reducing sugars[21,22]. This was further extended by Baly and his associates who were also able to show that UV of 200 nm wavelength could form, from CO_2 in the presences of fresh $Al(OH)_3$ or of other colloids, complex, gummy, carbohydrates that following acid hydrolysis, yielded reducing sugars[23]. Baly et al. had also shown, in 1922, that α-amino-acids could be generated by UV light acting on HCHO in the presence of potassium nitrite[24]. These remarkable experiments inspired J.B.S. Haldane to put forward his 1929 theory on the origin of life and on its precursor

Note [5]: It is quite impressive to read Butlerow's last line of his communication to the Academie des Sciences: "on peut dire que c'est le premier exemple de la synthèse totale d'une substance sucrée". "One may say that this is the first example of the total synthesis of a sugar".

hot, dilute soup. These "old English" experiments have been contested on the basis that small carbohydrate contaminants had been present in the formaldehyde used by them[25]. However, the careful exclusion of all carbohydrate contamination has clearly shown that UV irradiation of highly purified formaldehyde in the presence of the minerals alumina, montmorillonite or calcite produces a complex mixture of long —and branched— chain carbohydrates, similar to Baly's gummy product as well as the sugar pentaerythritol[26]. Equivalent results have been also reported by another Russian group, that of Pestunova et al[27]. Looking back in retrospect, the experiments of Butlerow and of the English chemists now appear as crucial for a modern interpretation of the origin of life. In the first place, the biotropic sea I have described contained abundant dissolved CO_2, as well as carbonic acid and bicarbonate. This weathered the surface anorthite that continuously released "fresh" $Al(OH)_3$, the same "impurity" that Baly found to be catalytically active in his experiments. Also, the UV light emitted by the sun at 4000 Myr b.p. had an 8-fold higher intensity than at present[28], and reached the Earth's surface with little decrease because of the absence of an ozone layer. The sea was strongly agitated by wind at the time, so its constantly renewed surface layers were exposed to this strong UV radiation. On the other hand, materials, not only $Al(OH)_3$ but a host of other mineral grains generated by rock fragmentation and comminution, would be kept in suspension and would favor the interaction of molecules previously synthesized both in the atmosphere and on land, now also present in the sea.

CARBOHYDRATE FORMATION

From the pre-1940 findings reported, we can hypothesize a mechanism for the prebiotic synthesis of sugars. It could have been as follows: UV irradiation of bicarbonate-rich sea, in the presence of $Al(OH)_3$ or other mineral particles, generated complex polysaccharides which, on irradiation, would produce formaldehyde that, in turn, in the presence of the also abundant Mg^{2+}, would be converted, also by UV irradiation, into glycolaldehyde and glyceraldehydes, that are potent initiators of the autocatalytic formose reaction that, in turn, is able to generate different monosaccharides, including ribose, one of the sugars present in nucleic acids[20,27]. Although experimentally the yield of ribose from this reaction was initially quite low (about 1%), and ribose was found to have a half life of only a few min at 80°C, recent experiments in which phosphates, pyrophosphate and borate catalysis have been used, have led to much higher yields of ribose and to its greater stability[29, 30, 32,33]. Also, this early ocean worked as a continuous chemical reactor in which any degradation of ribose could have been replenished by new synthesis, thus providing abundant ribose for eventual nucleic acid synthesis.

In summary, the prebiotic synthesis of the carbohydrates that living beings would need in the distant future is now, at least in principle, fairly well understood.

POLYPHOSPHATE: A KEY ELEMENT IN PREBIOTIC SYNTHESIS

The formation of polyphosphate (polyP) in KREEP minerals and, thus, in the biogenic sea where KREEP dissolved, produced a high concentration of this high-energy phosphate donor that could have a length of up to several hundred phosphoryl units, depending on the temperature of its formation. This polyanionic molecule was probably concentrated near the also anionic carboxyl groups of fatty acids, to which they were bound by Mg bridges. Such lipid association tended to place polyP in close apposition to liposomal membranes with part of its long tail sweeping both the intraliposomal space and the external medium where it could react with water-soluble molecules. The phosphorylation reaction rates could have been rather high because of the sea temperature and agitation[33].

In relation to lipid synthesis, phosphorylation of carboxylic (fatty) acids, of glycerol and of the amino alcohols choline, ethanolamine and serine (a hydroxylated amino acid) could allow with great ease and in large amount the synthesis of mono, di and triglycerides and phospholipids in an aqueous medium, without the unrealistic assumption of a dry beach.

Likewise, phosphorylation of ribose and of deoxyribose, as we shall see, could have led to the formation of the sugar-phosphate backbone of nucleic acids.

THE IMPORTANCE OF BEING PHOSPHORYLATED

The importance of the attachment of a phosphoryl group to an organic molecule, be it prebiotic or produced by a living organism, lies in the fact that the phosphoryl group is strongly electronegative, that is, it attracts electrons and this, through an inductive effect, makes the nearby carbon atoms electron-poor, i.e. with a partial positive charge, that makes them more susceptible to bonding by an electron-rich group of another molecule (i.e a nucleophilic attack), greatly increasing the chemical reactivity of these molecules as is clearly seen in the glycolytic pathway of carbohydrate metabolism.

CHAPTER 6

INFORMATIONAL MACROMOLECULES AND THE EMERGENCE OF CELLS

THE BIRTH OF RNA

The abiogenesis of nucleic acids raises a number of problems that were not found in the cases of lipid and non-genomic protein synthesis, as these occurred in an homogeneous aqueous medium. Nucleic acids, on the contrary, were formed from strongly hydrophilic polyP and ribose. The former, a mineral constituent and the latter, a product of the formose reaction that presumably occurred in the biogenic sea. The nucleic acid bases, in turn, were generated abiotically in the atmosphere and, later on, rained onto the Earth's surface profusely. Each of these bases is characterized by having a conjugated system of double bonds that makes them planar, UV–absorbing and highly lipophilic, so that, in the biogenic sea they partitioned into the lipid phases present, that is, they were in a phase that was separated from the aqueous medium where polyP and ribose were present. A hypothetical solution to this problem could be that ribose, in the case of RNA, was phosphorylated by polyP in its 3' and 5' carbons, leading to the formation of a ribose-phosphate chain with phosphodiester bonds joining the 3' and 5' carbons of vicinal riboses, like those seen in RNA. This "bottom up" configuration could solve the ribose instability problem. Also, the phosphorylation of ribose in the 3' position, as well as the OH group present in the 2' position, by an inductive effect, could also make the C-1 carbon electron–deficient and more prone to a nucleophilic attack by an electron–rich molecule. In the case of RNA synthesis, the nucleobases present in the lipid phase, by the absorption of UV photons, the flux of which was 8-fold greater than at present[1], could raise their electronic potential so that they could perform a nucleophilic attack on the C-1 carbon of ribose, thus allowing the final step in RNA synthesis. In turn, the potential energy of the phosphoribose moiety of RNA would also be enhanced by the high temperature of the sea water at the time, thus increasing its reactivity towards nucleobases. Once the coupling of the nucleobases to polyphosphoribose occurs, the marked hydrophilicity of ribose and phosphate would allow the final product (RNA) to remain in the aqueous phase where it could start to replicate.

Although this scheme presented for prebiotic synthesis of RNA is purely hypothetical, it nevertheless indicates that such synthesis could have occurred through this or some other feasible mechanism.

LIPOSOMAL GROWTH AND LATERAL TRANSFER OF GENETIC INFORMATION

The increase in the amount of RNA, due to increased replication, and, presumably also of proteins inside small liposomes, could have led to an increase in colloid-osmotic pressure inside them and to their fusion with other liposomes, thus creating progressively larger ones[2-4]. These fusions also led to the lateral transfer of genetic material, a mechanism that is thought important in the early stages of life[5].

RIBOZYMES: AN ESSENTIAL STEP IN CELL FORMATION

The discovery of RNAs with catalytic activity, the ribozymes, in which replication and enzyme activity, two of the essential features of living beings, are present, opened the way both for understanding how DNA was formed and how the first cells appeared[6-7].

We can now picture our biogenic sea as one in which billions of closed, liposomal–like structures, with abiotically formed proteins present inside them. It was in these minute pre-cellular structures where the synthesis of RNA began. Probably not synchronously nor with the same length or nucleotide sequences and, thus, with different enzyme capabilities. Each of these primeval RNAs would then start to replicate and the products formed, due to the high error rates characteristic of RNA replication[8], would start to deviate from the original sequence forming what Eigen and Schuster have named as a quasi–species[9], in which new functions appeared. These myriads of ribozymes could then be subject to Darwinian competition and selection, in which the selected characters would be those that increased the ribozyme replication rate and, thus, the formation of new and more abundant quasi–species. Among these newly formed RNAs one can conceive the emergence of transfer RNAs (tRNAs) in which activated amino acids could couple to one end of the molecule, while another region of this RNA base-paired to a certain region of a different RNA molecule. Variation of this primeval tRNA–like molecule could change the sequence of the base–pairing region that, if followed by a variation in the sequence of the ribozymic mechanism that couples an aminoacid to the terminal region of this primitive RNA, one can envisage this as the beginning of a very primitive RNA-based genetic code and of the first evolutionary mechanism leading to protein synthesis. Crick has presented a preribosomal mechanism of protein synthesis that has a certain similarity to this one[10,11]. Also, as it is known that the activity of peptidyl-transferase, the enzyme that forms peptide bonds on the ribosome, is due to a ribozyme[12], it may be thought that it was also active in precellular structures, forming RNA coded proteins, amongst them ribosomal ones, thus leading to the formation of primitive ribosomes.

FROM RNA TO DNA

There are several indications suggesting that RNA was a precursor of DNA. Among them is the current need for an RNA primer in the synthesis of DNA by DNA polymerase[13]. Also, obviously, RNA is needed in RNA-dependent DNA polymerase, or reverse transcriptase, an enzyme first found in the replication of some RNA viruses, such as that of HIV, that inserts the RNA viral sequence into that of the host's DNA. This was later also found to operate in the eukaryotic enzyme telomerase[14] and in the transposition of DNA segments (the transposons) via an RNA intermediate[15]. This latter is known to have a very ancient evolutionary origin and may have operated in precellular structures[16,17].

Chemically, the formation of DNA from RNA is relatively straightforward. First, ribose must be reduced to deoxyribose, as now occurs by the action of ribonucleotide-reductase. Then thymine must be generated because this nucleobase has an electronic potential that is high enough to make a covalent bond with the C-1 carbon of deoxyribose, as already indicated. Before the generation of thymine, only DNA containing cytosine (C) and guanine (G) could be possible, as uracil is unable to make a bond with deoxyribose. This has also been suggested by the predominance of G and C in what is thought to have been the most primitive genetic code[11]. A DNA rich in G and C is also well suited for the DNA stability at the high temperatures present in the biogenic sea, as G forms three hydrogen bonds with C in the double helix, against only two in A-T pairs.

It seems possible that in the origin of life a mechanism similar to that of reverse transcriptase operated, be it due to a ribozyme or to a protein enzyme, by which the nucleotide sequence of RNA was converted to the same sequence in DNA. This would have been a giant leap in the origin of life, as the sequences of the RNAs that had been more successful in their Darwinian competition now became incorporated into a molecule that is chemically much more resistant to hydrolysis (because of the absence of the 2' OH in deoxyribose) and because DNA has a copying error five or six orders of magnitude lower than RNA (10^{-10} to 10^{-12} in DNA versus 10^{-4} to 6×10^{-5} in RNA, both expressed as errors per replication round and per nucleotide)[8].

AND THEN THE FIRST CELLS

The synthesis of informational molecules in the primitive liposomal-like compartments must be thought as occurring in a large population of partially independent compartments. In the case of RNA, the building materials (ribose, polyphosphate and nucleobases) were present probably in most of the compartments. In the case of DNA this was not the case, as deoxyribose was derived from ribose, a process that was not automatic, and thymine probably also had a secondary origin. This could indicate that not all RNA molecules produced DNA, and that this could appear only in restricted zones of the liposome world. From these islands DNA could multiply, colonizing neighbouring regions. As DNA presented a great selective advantage as an information carrier over RNA, most of the functional features that had formed in the primitive RNA, were transferred to DNA, and Darwinian selection now became one in which DNA was the almost exclusive player. In this selection process not only linear descent operated but also lateral transfer of DNA molecules was important[5], as well as the joining of them in larger and more complex ensembles, finally leading to molecules with sufficient information content to sustain the basic functions of an independent living cell[18]. These were a replication mechanism for DNA, a transcription mechanism for the synthesis of the different types of RNA and a protein synthesizing machinery that utilized the genetic code, at least a primitive version of it, for the formation of gene-encoded proteins.

It is not known whether the first cell was a single one or a family of closely related structures from which all other cells, in the course of several billion years, descended. As to when the first cell(s) emerged, the oldest known microfossils have been dated at 3500 Myrs b.p.[19] Although morphological studies have been debated[20], new molecular studies made on these fossils do support old age[21-23], while molecular studies of transfer RNA indicate a maximal age of 3800 Myrs. This could be, thus, the time when life on Earth began[24].

In the early stages of life many cell materials were preformed, i.e. amino acids, lipids, including plasma membranes, nucleobases, sugars, etc., and a powerful energy source, in the form of polyP, was also present. There was, therefore, no need for mechanisms that could synthesize them. Only after one of these components dwindled, a selective pressure for their formation appeared. The same may be said when the ionic composition of the biogenic sea started to change, probably because of mixing with a sodium – and calcium– rich ocean. In this case mutants that were able to extrude sodium and calcium from the cells by means of energy–dependent ion pumps, had to be generated.

We have followed the origin of the first living cell from what De Duve called "vital dust"[25], that is, cosmic grains, to the formation of our solar system and of the Earth. And then to many changes on the Earth's surface and atmosphere, to the generation of a huge wealth and variety of biochemical compounds that became more and more complex in what I have called the biogenic sea, an essential element in the origin of life and, finally, to the extraordinary self organization of these biochemical molecules leading to a first living cell. In all these steps, including the physical and chemical ones, selective mechanisms that can be equated to Darwinian selections, were present.

Many of the steps leading to the origin of life, especially the early physical and chemical ones, may be deduced with a reasonable amount of certainty. However, the more recent biochemical processes that occurred in the biogenic sea are, in spite of the numerous findings published during the last decades, largely conjectural and speculative.

As Eigen stated: "We know in many cases how things could have taken place, but not how they did take place"[26], and despite so many efforts, what Darwin expressed in a letter to G.C. Wallich: "That he had left the origin of life uncanvassed as being altogether ultra vires (beyond man's capacity) in the present state of our knowledge"[27], still holds true in many respects. Nevertheless, Darwin's foresight on this subject is remarkable as shown in his letter to J.D Hooker of February I, 1871: "But if (and oh! What a big if!) we could conceive in some warm little pond, with all sort of ammonia and phosphoric salts, light, heat, electricity, etc., present, that a proteine [note 6] compound was chemically formed ready to undergo still more complex changes, at the present day such matter would be instantly absorbed, which would not have been the case before living creatures were found". This, according to our hypothesis on the origin of life could be translated as follows: "The warm little pond" is our warm biogenic sea; "all sort of ammonia salts": amino acids and nucleobases; "phosphoric salts": polyphosphate; "a proteine" compound ready to undergo still more complex changes: abiotically formed proteins and nucleic acids that were yet to be discovered. "Instantly absorbed": biological molecules were degraded by proteolytic and other hydrolytic enzymes produced by living cells.

Note [6]: Darwin used the word proteine (from the French *protéine*).

PART TWO

THE ORIGIN OF LIFE
IN PERSPECTIVE

"How strange that things exist".

L. Wittgenstein

CHAPTER 7

CHANCE, SPONTANEITY AND INFORMATION

THE ORIGIN OF LIFE ON EARTH: A SPONTANEOUS PROCESS?

We have followed the history of life starting with the origin of the Solar System, showing a succession of causally related events. First the presence of cosmic dust granules, originated in the shattering of ancient supernovae. Among them carbon—rich stars from which black carbon-containing granules derived. These were responsible for the black colour of molecular clouds in which our sun and planetary system were born. These myriads of tiny black grains absorbed the prevalent intense electromagnetic radiation that would have destroyed any structure formation since its inception. This allowed the formation of the Solar System.

Next, the agglomeration of dust granules with their mineral cores and shell of frozen gases, including water, CO_2, methane, NH_3, CO, H_2CO and others necessary for life, gave rise to the Earth and the other planets, while hydrogen, the most abundant element in the universe, started its gravitational compression and temperature rise in the center of the galactic disk, generating the sun, in which fusion of H nuclei generated the light and heat in which life could thrive.

The Earth, formed from the accretion of dust granules and of progressively larger bodies, heated and melted into an igneous sphere in which the denser elements, such as iron and nickel, sank, forming a hot dense core, while the lighter elements floated originating the crust and mantle. The presence of a metal core generated a strong magnetic field that deflected the energetic solar wind. This was an essential feature for the protection of life. Next, the catastrophic event in which the Moon originated, incremented the iron core of the Earth and, thus, the protection from the solar wind. Also the Moon-forming impact increased the mass of the Earth and its distance from the Sun, placing our planet well into the "habitable" zone, where liquid water and moderate temperatures could exist. Also the Moon stabilized the Earth's axis and made possible the existence of seasons.

The Moon-forming impact rewarmed the Earth. This generated a second atmosphere formed by hot water vapor, CO_2, CO, CH_4, COS, NH_3 and other gases that were initially present in the core-mantle cosmic grains and had remained occluded inside the Earth. It was in this new atmosphere, to which H_2 was later added, in which high energy lightning bolts induced the formation of a large amount and variety of organic molecules, including amino acids, nucleobases and fatty acids, that would be essential for the formation of living beings.

In a period that lasted from 4100 to 3850 Myr b.p. a remarkable event occurred: the giant planets Jupiter, Saturn, Uranus and Neptune changed their orbits, with Neptune migrating outwards and dispersing 97% of the mass of the planetesimal disk. These events led to a prolonged asteroid bombardment of the inner Solar System, including the Earth, on the surface of which many depressions were formed, some of which would be later filled by the condensation of the steam atmosphere, giving rise to lakes, seas and oceans.

One of the important effects of this bombardment was the excavation of the Earth's crust, exposing deeper mineral layers that would contribute to the ionic composition of the future "biogenic" sea, amongst them KREEP, the mineral that would provide the potassium needed for life, as well as polyP, the primordial chemical energy source, responsible for virtually all of the biochemical synthetic reactions that would take place in this primordial sea, leading to the formation of biomembranes, proteins, RNA, DNA, and eventually, to the first living cells.

In the atmosphere, the intense heat generated by the falling asteroids, together with the effect of lightnings, generated vast quantities of
H_2, making it a highly reducing medium necessary for the formation of
organic compounds.

When the Earth cooled once more, so did its atmosphere. Then the
steam atmosphere started to condense giving rise to copious rains that
filled the depressions excavated by the meteor bombardment. This created seas, lakes, etc. The rains also washed the organic molecules that
were present in the atmosphere, part of which dissolved in the biogenic
sea. This hot rain weathered the mineral anorthite present in the early
crust and led to the drastic reduction of atmospheric CO_2, decreasing
the absorption of outgoing infrared radiation (the greenhouse effect),
thus lowering the Earth's temperature and favouring its habitability.

In the biogenic sea the formation of sugars by the formose reaction and
the non-genomic synthesis of polypeptides seem well founded. However,
the synthesis of more complex lipids, of RNA and of DNA is more conjectural and speculative, so the precise steps leading to them remain undefined until more experimental evidence is provided. However, it cannot be
doubted that, the same as previous steps in the origin of life, they were due
to the spontaneous operation of chemical and physical forces.

In the brief summary presented, we have seen a long list of events
leading to the origin of life, and these events, seen in a limited perspective, all appear as spontaneous. But this must be examined further.

DID LIFE ORIGINATE BY CHANCE?

The most classical example in the study of aleatory (chance) events is
that of coin tossing. Given a sufficient number of tossings, the number
of "heads" obtained will approach 50%, and so will the percent of "tails".
However, if we build a coin-tossing apparatus, free of air currents, in
which exactly the same impulse is given to exactly the same place on the
coin, then the outcome of 100% of the tosses shall be "heads" (or the

opposite). In my opinion, this indicates that chance is only an ignorance of causes, not a suspension of the principle of causality, and that Darwin's opinion (in a letter to W. Graham of July 3, 1881): "My innermost conviction" is that "the universe is not the result of chance", is certainly valid.

Natural Law and the Origin of Life

In our description of the many events that took place leading to the origin of life, all may be, one way or another, reduced to the interaction of different parts of the material universe: physical interactions such as those due to gravitational or electromagnetic forces, to the different types of nuclear forces or to thermal effects; and chemical interactions, as a consequence of the electronic structures of different atoms, that lead to the formation and breaking of covalent bonds, or to ionic or hydrogen or other types of less stable bonding.

One may consider that spontaneity is the undisturbed operation of a system according to natural laws. However, natural laws are not abstract entities but, rather, they are ingrained in the structure of the material elements present in our Universe. Thus, natural laws are a consequence of the very nature of these elements, and spontaneity, therefore, depends on the origin of such elements, as we shall see.

Spontaneity and Information

We have seen that the first living organism appeared because of a very long chain of events that occurred in our Universe during many millions of years. If one examines each individual event, one must conclude that it was of spontaneous occurrence. But, if one looks in perspective at the whole chain of events, it emerges that the exact coincidence and timing of such different causes was not casual or aleatory, but that they cooperated to achieve a final goal: the emergence of life. Thus, this process contains information. We can compare the situation to a canvas that is being painted by an artist. Each time the artist paints a stroke on it, the paint is transferred to the canvas by a number of spontaneous physico-chemical mechanisms. However, when the painting is seen in the proper perspective, information or design become apparent. This is quite analogous to the origin of life, where information content may be seen only when the proper perspective, both temporal and causal, is used to examine the multiple causes that contributed to it.

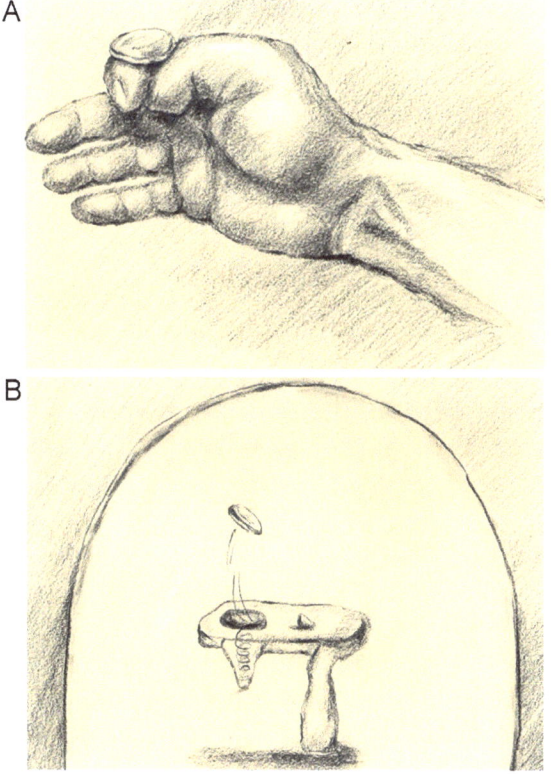

***Figure 15:* COIN TOSSING**

In a: hand tossing of coin is shown; in b., a coin-tossing machine, free of air currents, that gives the same coin always the same impulse in the same place. In a. the results approach 50% for heads or tails; in b. results will be close or equal to 100% heads (or tails). (Courtesy: Catherine Ocqueteau).

WHAT IS LIFE?

In his famous book "What is Life?"[1], Erwin Schrödinger was impressed by the fact that, contrary to what is seen in physical systems, where disorder (i.e. entropy) tends to increase until an equilibrium is reached, or, what is the same, entropy is maximized, living systems exist far from equilibrium and equilibrium is reached only with death. Because of this Schrödinger postulated that living creatures had to balance entropy increase by means of "negative entropy". He thus stated that living organisms "feed on negative entropy", the nature of which remained undetermined. However, when we examine the probabilistic formula of entropy engraved on Boltzmann's tomb:

$$S = k \log W$$

where S is entropy; k, Boltzmann's constant (3.2983 by 10^{-24} cal$/°$C), and W the probability of the system (or, more specifically, the number of possible microstates of that system), we can deduce from it that negative entropy relates to the amount of information [note 7]. The equivalence of negative entropy and Shannon's information content was first deduced by Leon Brillouin (1889-1969) and is expressed in the formula $\Delta S = -k \log_e 2 \, \Delta H$, where ΔS is the change in entropy and ΔH is the change in Shannon's information content[2]. It should be noted that this is not a relation of identity but of equivalence, in which Boltzmann's constant k is the conversion factor. Or, in other words, that we can replace Schrödinger's idea by saying that living beings "feed on information", or that living beings need information to maintain their far-from-equilibrium state, and the information, provided especially by nucleic acids and, in a much lesser scale, by

---

Note [7]: $S = k \log W$; $\frac{S}{k} = \log W$; $-\frac{S}{k} = -\log W$, and $-S = k \log\frac{1}{W}$, where $\log\frac{1}{W}$ is the reciprocal of disorder, i.e. order, a consequence of the information content of the system.

proteins, is able to order the flow of matter and energy in the steady state we call life.

Schoenheimer showed that in mammals the body constituents are in a dynamic state[3], that is, its material components, e.g. proteins, lipids, carbohydrates, etc., with few exceptions, are continuously being replaced by similar or identical molecules, but the form of the organism does not change. This stability of the overall structure, in spite of material flux, is maintained by the flow of information. It is similar to the idea of the river of Heraclitus (544 – 484 B.C) who said, that because everything flows (including the river's water) no one can bathe twice in the same river, and to G. Buffon's (1707 – 1788) idea of something that kept constant the form of the body (i.e. information) was due to an "internal mould", which we now know is DNA.

The idea that multicellular living beings are continuously building themselves, led to Maturana and Varela's influential concept of autopoiesis[4]. However this self-construction process is valid only for higher organisms and only under certain circumstances. In bacteria, the most numerous of living creatures, each individual bacterium uses its DNA-encoded information to organize nutrients into a larger cell and, eventually, to a daughter cell, in a process known as binary fission. Thus, what occurs is not a construction of self (autopoiesis) but a construction of a similar but different being, and this new cell is many times subject to variation, so living beings are really in an process of information-directed "homoiopoiesis" (from the Greek omoio -ὅμοιος-, similar, and poiesis -ποίησις-, to make). In fact, multicellular organisms not only generate themselves but also generate descent by means of reproductive mechanisms, that is not an autopoietic process but, rather, an alteropoietic one. This concept of homoiopoiesis thus englobes that of autopoiesis but extends it to the generation of similarity and, thus, to that of reproduction and of genetic variation, both essential attributes of living beings.

An Extraterrestrial Origin?

It has repeatedly been argued, even by people as bright as Fred Hoyle, that life on Earth had an extraterrestrial origin. This has been based on the relatively short life of the Earth, on the extremely low probability of the assembly of a living cell from its component parts, and on the widespread presence of organic molecules in the Universe. But, although the age of our planet is low (about 4500 Myrs since the Moon-forming impact) for complex living beings to form "spontaneously", the duration of the Earth has been only one third of that of the Universe, estimated as 13,820 Myrs starting with the Big Bang, that is, the time difference in itself is not impressive. Also, although many life-related organic compounds are indeed present in the Universe, they are much too dilute to interact among themselves. In meteorites that reach the Earth, there are also many organic molecules, however, the high temperatures that affect them during their entry into the Earth's atmosphere and also during their impact onto the surface, would not allow any living organism to survive.

The probability of molecular self-assembly into a living cell, however simple this may be, is vanishingly low[5]. This validates the argument of insufficient time for the generation of life on Earth by chance mechanisms, however, it does not argue either in favour of its extraterrestrial origin, as time in the Universe seems equally insufficient. It argues, rather, for the impossibility of the generation of life by mere chance, except, as it is postulated in this book, if an informationally-directed mechanism has operated.

CHAPTER 8

THE ORIGIN OF LIFE AND DARWIN'S CONCLUSION

LIFE AND MATTER

Because we now know that life is a unique form of informationally organized matter, we must also ask: What is matter? Although we ignore what the ultimate material element is, we can say that the most salient feature of matter is its discontinuity, i.e. it is formed by limited parts, e.g. galaxies, rocks, molecules, atoms, subatomic particles, energy quanta, etc. and, however large the number of discrete parts, it is always limited because each discrete part has its own limits.

We also know much of the genesis of matter in the early Universe, with an energy-dominated Universe at first, formed by extremely high-energy quanta that evolved towards the formation of quarks and antiquarks, and then to particles and antiparticles, followed by a selective process in which particles predominated over antiparticles, until the first atomic nuclei, the heavy isotopes of hydrogen (deuterium and tritium), helium and lithium formed a few minutes after the Big Bang, while the hydrogen nucleus (a single proton) was formed earlier. Much later, all the rest of the chemical elements, including those necessary for life, were formed inside large stars that ended as supernovae, seeding the Universe with their precious products. In all these processes natural selection mechanisms operated with survival of the fittest, (or more stable), elements may be seen. This is particularly evident in the nucleogenesis in stars, in which fusion of atomic nuclei have different outcomes: From fleetingly brief products, to unstable radioactive elements and to fusions that produce stable elements.

It is not known whether a pre-Big Bang material world existed, from which our Universe and other possible ones derived, but if it did, this "protomatter" must have had a certain homology with the matter we now know (it was a Darwinian common ancestor). If not, our present matter could not have derived from it. In the Universe there is a sort of genetics in which one element derives from a pre-existing one, e.g. subatomic particles derive from quarks, and the nuclei of different chemical elements derive from hydrogen (a proton), which generates the progressively larger nuclei of deuterium, helium, berillium, carbon, nitrogen, oxygen and all the other natural elements present in the Universe. Therefore, this hypothetical protomatter must also have had a discontinuous and physically limited character. If this pre-Big Bang protomatter gave rise to our Universe or, to other universes, given its physically limited character, generation of new physical entities should have depleted it, at least in part: it was not an inexhaustible reservoir and could not have an infinite duration.

Regarding our Universe, we must conclude that it is spatially limited, just as the matter that forms it: It has internal and external limits. We can also say that our Universe is dynamic: it changes in time and the changes are basically irreversible. This is due to the nature of entropy that, if uncoupled from informational processes, cannot decrease spontaneously. For instance, if a glass of water is turned on a table, the spilled water molecules will never return to the glass, and in a star, the fusion of four atoms of hydrogen into one of helium, with the release of electromagnetic energy, positrons and neutrinos, cannot be reversed with radiation and particles coming back to the helium, and reconstituting the four H atoms. This is known as the "arrow of time"[1-3].

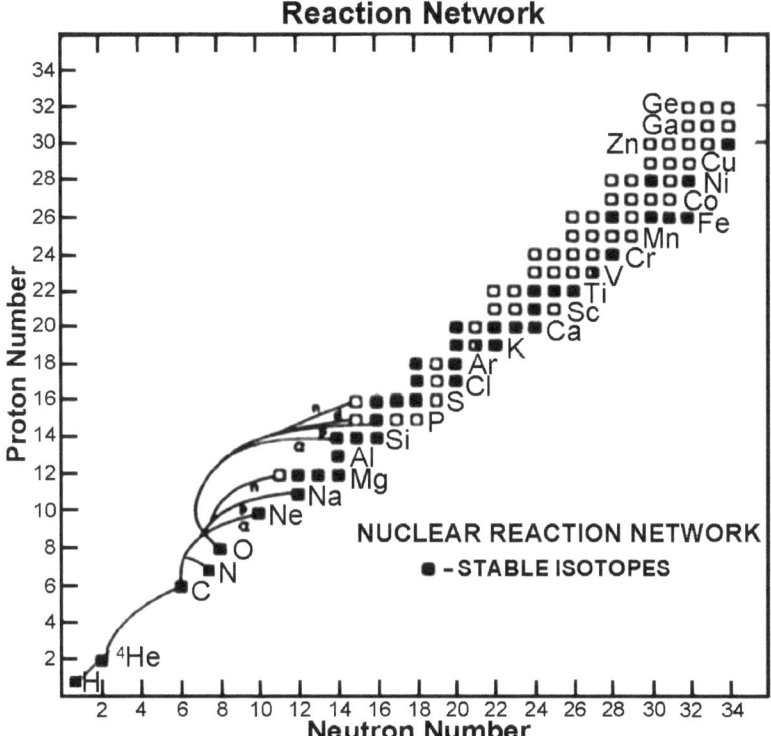

Figure 16: FORMATION OF NEW CHEMICAL ELEMENTS IN THE INTERIOR OF STARS
In the hot interior of stars only free atomic nuclei, stripped of their electrons, are present. When the temperature is high enough to overcome the electrostatic repulsion of the positively charged protons present in nuclei, the fusion of smaller nuclei gives rise to larger ones. Initially, stars use hydrogen (H) as fuel, with release of energy when hydrogen nuclei fuse to form helium (He) (See Figure 4). When the H supply dwindles, He fusion shall follow, giving rise to beryllium, and then, by the fusion of three He nuclei, carbon, the fundamental building block for life. When the stars are large enough, this series of fusions will continue up to Fe and Ni, after which the fusion process is unable to generate energy and the star rapidly collapses under its own gravity. This is followed by an explosive rebound with the formation of a supernova. This explosion generates an enormous amount of energy that enables further nuclear fusions and the generation of all the chemical elements heavier than iron and nickel. Some of the elements formed along these processes are stable (solid squares), while others are unstable (radioactive) and shall disintegrate (open squares). (Source: Woosley, Arnett and Clayton, 1974, modified).

Stars in our Universe use fuel: first hydrogen, then helium, and, if large enough, carbon, oxygen, silicon and heavier elements up to iron. The thermodynamic principles in stars are the same as those that operate in an oil lamp. The lamp shines while it has oil and will stop doing so when all its fuel is exhausted. The fact that there are shining stars at present indicates that they had a beginning, that their duration is not infinite. It has been estimated that all stars will be dead in 10^{23}yrs. It has also been estimated that in a flat or open universe, given enough time, black holes will evaporate, protons will decay and only electrons, positrons, neutrinos, antineutrinos and gamma rays will exist[4].

Thus, our Universe is limited, both spatially and temporally and, independent of any particular model, such as the Big Bang, material universes are also limited in time: They have a beginning and end in a state of maximal entropy or almost complete disintegration.

Einstein's Greatest Blunder, Dark Energy and Friar Ockham

When in 1917 Einstein published his General Relativity Theory he was convinced, as was the case with all scientists then, that the Universe was static and did not change with time[5]. But, as Einstein noted that with the equations in his theory, gravitation would lead to the collapse of the Universe, he introduced a new force which he named the cosmological constant or lambda. It would be exactly the opposite to gravity: It would be repulsive and would cancel the pull of gravitation, allowing the Universe to remain static[5]. In 1922 the Russian mathematician Alexander Friedmann realized that lambda was unnecessary and that the Universe could be either expanding or contracting[6,7]. After some time Einstein accepted, reluctantly, this modification and called his cosmological constant "the greatest blunder of my life"[8].

Recently astronomers studying the velocity of recession of distant stars, using the brightness of type Ia supernovae as "standard candles", have found that distant stars, instead of reducing their velocities by the pull of gravity, are increasing it. This outward pull has been thought to be due to a mysterious repulsive force nicknamed dark energy, that acts as an antigravitatory force[9,10]. Hélas! Einstein's cosmological constant has resurrected.

One of most recurrent ideas in Humanity has been the belief that our habitat occupies a central position: the Earth, in the long held Ptolemaic cosmological model. Then came Copernicus' heliocentric model. Later on, it was thought that our galaxy, the Milky Way, was the sole one. This belief subsisted until the studies of Edwin P. Hubble on the twinkling Cepheid stars showed that many more galaxies were present[11]. Now it is implicitly thought that our Universe is the only one that exists. But this may not be so, and a simpler interpretation of the nature of dark energy, following the parsimony principle of William of Ockham (1280 – 1349), that states that given several explanations for a given phenomenon, it is better to chose the simpler one. In our case, if we postulate that "our" Universe is embedded in a much larger one, where our well known gravitation is generated, it would appear that the objects in our Universe, especially the more distant ones, are subject to a "strange" repulsive dark energy that increases with distance. But this would only be reflecting the gravitational attraction exerted by the larger Universe in which we are embedded. If this explanation were true, and an Ockham's principle is not a certificate of truth, it could mean that our universes are similar to Russian dolls, where smaller ones are contained inside larger ones, an idea that Blaise Pascal (1623 – 1662) with his worlds inside worlds, inside worlds, would have certainly liked[12].

The possible existence of multiple universes raises the possibility that the Creator could have produced many types of "most beautiful and most wonderful" living beings that go beyond our imagination.

GOD, TIME AND LIFE

We have seen that the material Universe had a beginning, but not how far back in time this happened. We also now know that, contrary to what was believed for centuries, there is no absolute space and time but, after Einstein's General Relativity it is accepted that both space and time form part of the Universe, they are "generated" by its dynamical material elements. In philosophical terms they are relative accidents, they do not have an autonomous existence but exist in relation to something and are different from absolute accidents that "inhere" in a substance. Saint Augustine (354 – 430) had already grasped this relativity of time and its dependence on the Universe when he stated that "God created time that started to exist together with heaven and earth"[13]. In turn, Ludwig Wittgenstein reflected: "The solution to the riddle of life in space and time lies outside space and time"[14].

The issue of a beginning of matter and time immediately raises the question of how this could have began from non-matter or, more specifically, from absolute nothingness. To answer this, as we are now in a unique situation in which nature and, thus, natural laws, still do not exist, we must turn to philosophy, to a theoretical-rational approach.

The first answer came from Parmenides (530? – 470 B.C), who stated that "nothing comes from nothing", an undisputable, self-consistent idea. One step forward was given by Thomas Aquinas (1225 – 1274) who stated: "As active potency must be more perfect when the passive potency is lesser or more imperfect. When passive potency is reduced to nothing, active potency must be infinite. Only God can create something from nothing"[15].

The same conclusion was reached by D. Atkatz and H. Pagels in 1982: "We hoped that the Universe could have originated (by a quantum tunnelling event) from flat empty space, a configuration that truly corresponds to nothing at all. Unfortunately, this possibility seems to be ruled out by the lowest energy state. A deviation from such a geometry must be paid for by a corresponding increase in the action, and changing the action over an infinite volume costs infinite energy"[16].

The conclusion of this is that the Universe and all its components, including living organisms, cannot have derived spontaneously from absolute nothingness and a cause that is external to material universes has to be sought for.

From the knowledge of the material created Universe we can deduce some characteristics of the entity that created it: First, that it exists. Secondly, that it is independent of space and time, that form part of the created Universe. It is, therefore infinite (extra spatial) and eternal (independent of time). Thirdly, creation from nothing reveals an infinite power. Fourth, the information content leading to the dynamic ordering of many different elements ending in countless structures, including living beings, reveals an extraordinary intelligence. An entity with such characteristics has generally been called God.

ON CREATION, CHAOS AND BUTTERFLIES

One important new scientific paradigm that appeared in the 20th century is that of Chaos theory[17]. It has become popular under the name of the "butterfly effect", based on weather prediction studies by Edward Lorenz in 1963[18,19]. It states that small changes in initial conditions lead to large differences in future outcomes, so that, although the events are deterministic, the outcome is unpredictable. However, the more perfect the knowledge of causes (e.g. an increase in the number of stations that measure climate), the better the predictability of a given phenomenon. In the case of the origin of the Universe, no human mind is able to grasp it in all its complexity and, therefore, unable to predict its future evolution. Only the mind of God was able of such knowledge, so that the future evolution of the Universe was set at the very instant of its creation and, due to its sensitivity on initial conditions, we may say that the lack of a single quantum of energy in creation, would have led to an entirely different Universe, and the origin and evolutionary transformation of living creatures was a consequence of those initial conditions.

In nature, therefore, there is both creation by a first cause, and evolution by the spontaneous but informationally-directed operation of natural laws (secondary causes). In the hierarchy of things, creation is more important than evolution because it precedes and conditions it.

The conclusions we have reached, based on solid scientific data, are entirely in concordance with those reached by Charles Darwin in 1859 in the last chapter of "The Origin of Species" where he stated: "To my mind it accords better with what we know of the laws impressed on matter by the Creator, that the production and extinction of the past and present inhabitants of the world should have been due to secondary causes, like those determining the birth and death of the individual. When I view all beings not as special creations, but as the lineal descendants of some few beings which lived long before the first bed of the Silurian System was deposited, they seem to me to become ennobled."

"There is grandeur in this view of life, with its several powers, having been originally breathed into a few forms or into one;... from so simple a beginning endless forms most beautiful and most wonderful have been, and are being, evolved"[20].

Figure 17: **CHARLES DARWIN**
Charles Darwin in 1854, near to the time (1859) when he published "The Origin of Species".

REFERENCES

Chapter 1. FROM COSMIC DUST TO OUR SOLAR SYSTEM

1. A. Bouvier, M. Wadhwa. The age of the Solar System Redefined by the Oldest Pb–Pb Age of a Meteoritic Inclusion. Nature Geoscience 3:637-641 (2010).

2. P. C. Frisch. The Solar Galactic Environment. Physics of the Outer Heliosphere. AIP Conference Proceedings, 719:404-411 (2004).

3. P. C. Frisch, J. D. Slavin. The Chemical Composition and Gas-to-Dust Mass Ratio of Nearby Interstellar Matter. Astrophys. J. 594:844-858 (2003).

4. E. Herbst, E. F. van Dishoeck. Complex Organic Interstellar Molecules. Ann. Rev. Astron. Astrophys. 47:427-480 (2009).

5. J. Lequeux. The Interstellar Medium. Springer, Berlin, 2005; pp 209-226.

6. S. Messenger, S. Sandford, D. Brownlee. The Population of Starting Materials Available for Solar System Construction. Meteorites and the Early Solar System II, D. S. Lauretta and H. Y. McSween Jr. (eds.), University of Arizona Press, Tucson, 943 pp., p.187-208 (2006).

7. L.B. d'Hendecourt, L. J. Allamandola, J. M. Greenberg. Time Dependent Chemistry in Dense Molecular Clouds. I-Grain Surface Reactions, Gas/Grain Interactions and Infrared Spectroscopy. Astron. Astrophys. 152:130-150 (1985).

8. A. Li, J. M. Greenberg. A Unified Model of Interstellar Dust. Astron. Astrophys. 323:566-584 (1997).

9. J.M. Greenberg. The Secrets of Stardust. Sci. Am. 283, 70-75 (2000).

10. J.M. Greenberg, C. Shen. Cosmic Dust in the 21st Century. ArXiv: astro-ph/0006337v/ 23 jun 2000.

11. L. Dunne, S.J. Maddox, L. Rudnick et al. Cassiopeia A: a Dust Factory Revealed by Submillimetre Polarimetry. Mon. Not. R. Astron. Soc. 394:1307-1316 (2009).

12. J. Bally, J. Morse, B. Reipurth. The Birth of Stars: Herbig-Haro Jets, Accretion and Proto-Planetary Disks. Science with the Hubble Space Telescope – II. Book Editors: P. Benvenuti, F. D. Macchetto, and E. J. Schreier. Electronic Editor: H. Payne.

13. T. Montmerle, J-C. Augereau, M. Chaussidon, et al. 3. Solar System Formation and Early Evolution: The First 100 Million Years. Earth, Moon and Planets 98:39–95 (2006).

14. P.J. Armitage. Observation of Planetary System. In Astrophysics of Planet Formation. Cambridge University Press. (2010).

15. P.J. Armitage. Dynamics of Protoplanetary Disks. Ann. Rev. Astron. Astrophys. 49:185–236 (2011).

16. S.R. Cranmer. Turbulence-Driven Polar Winds from T Tauri Stars Energized by Magnetospheric Accretion. Astrophys. J. 689:316–334 (2008).

Multimedia Support
Youtube. "Birth of the solar system". National Stem Center. (3.05 min.).

Chapter 2. A New Earth and a New Moon

1. P. Goldreich, W. R. Ward. The Formation of Planetesimals. Astrophys. J. 183:1051-1062 (1973).

2. V.S. Safronov. The Protoplanetary Cloud and its Evolution. Soviet Astronom. 10: 650-658 (1967).

3. B. J. Wood, M. J. Walter, J. Wade. Accretion of the Earth and Segregation of its Core. Nature 441:825-833 (2006).

4. A. N. Halliday, B. J. Wood. How Did Earth Accrete? Science 325:44-45. (2009).

5. O.Yu. Schmidt. Four Lectures on the Theory of the Earth's Origin I₃d-w AN SSSR, Moscow, 3rd ed. (1957).

6. Kleine, C. Münker, K. Mezger, H. Palme. Rapid Accretion and Early Core Formation on Asteroids and the Terrestrial Planets from Hf–W Chronometry. Nature 418:952-955 (2002).

7. E.N. Parker. Dynamics of the Interplanetary Gas and Magnetic Fields. Astrophys. J. 128:664-676 (1958).

8. B.E.Wood, H-R. Müller, G.P. Zank, J.L. Linsky. Measured Mass Loss Rates of Solar-Like Stars as a Function of Age and Activity. Astrophys. J. 574:412–425 (2002).

9. H. Jenny. Factors of Soil Formation. Dover publications, INC. (1994).

10. W.K. Hartmann, D.R. Davis. Satellite-Sized Planetesimals and Lunar Origin. Icarus 24:504-514 (1975).

11. A. G. W. Cameron. The Impact Theory for Origin of the Moon. Origin of the Moon. Proceedings of the Conference, Kona, HI, October 13-16, 1984. Edited by W. K. Hartmann, R. J. Phillips, and G. J. Taylor, p. 609.

12. R. M. Canup, E. Asphaug. Origin of the Moon in a Giant Impact Near the End of the Earth's Formation. Nature 412:708-712 (2001).

13. A.N. Halliday. A Young Moon-forming Giant Impact at 10-110 Million Years Accompanied by Late-stage Mixing, Core Formation and Degassing of the Earth. Phil. Trans. R. Soc. A. 366: 4163-4181 (2008).

14. R. M. Canup. Dynamics of Lunar Formation. Ann. Rev. of Astron. Astrophys. 42:441-475 (2004).

Multimedia Support
Youtube. "The Birth of the Moon", Space Rip. (24 min.).
Youtube. "Kaguya Taking 'Full Earth Rise'". (1.15 min.)

Chapter 3. On Lightnings and Bombs

1. S.L. Miller. A Production of Amino Acids Under Possible Primitive Earth Conditions. Science, 117:528–529 (1953).

2. S.L. Miller and H.C. Urey. Organic Compounds Synthesis on the Primitive Earth. Science, 130:245–251 (1959).

3. A.P. Johnson, H.J. Cleaves, J.P.Dworkin, D.P. Glavin, A.Lazcano, J.L. Bada, The Miller Volcanic Spark Discharge Experiment. Science, Supplementary online material. 322: 404 – 404 (2008) Plus.

4. L. Schaefer, B. Fegley, Outgassing of Ordinary Chondritic Material and Some of its Implications for the Chemistry of Asteroids, Planets and Satellites. Icarus 186:462-483 (2007).

5. K. Zahnle, L. Shaefer, B. Fegley. Earth's Earliest Atmospheres in The Origins of Life. D. Deamer, J.W. Szostak (eds). Cold Spring Harbor Lab. Press, N.Y. (2010), pp.49-66.

6. L. Schaefer, B. Fegley. Application of an Equilibrium Vaporization Model to the Ablation of Chondritic and Achondritic Meteroids. Earth, Moon, Planets. DOI 10.1007/s11038-005-9030-1

7. J.I. Lunine. Physical Conditions on the Early Earth. Phil. Trans. R. Soc. B. 361:1721–1731 (2006).

7a. P. Hartogh, D.L. Lis, D. Bockelée-Morvan et al. Ocean - like water in the Jupiter-family Comet 103P/Hartley 2. Nature 478: 218-220 (2011).

7b. A.E. Saal, E.H. Hauri, J.A. Van Orman et al. Hydrogen Isotopes in Lunar Volcanic Glasses and Melt Inclusions Reveal a Carbonaceous Chondrite Heritage. Science DOI 10.1126/Science 1235142. May 9,2013.

8. R. Gomes, H.F. Levison, K. Tsiganis, A. Morbidelli. Origin of the Cataclysmic Late Heavy Bombardment Period of the Terrestrial Planets. Nature 435: 466 – 469 (2005).

9. W.F. Bottke, D. Vokrouvhlicky, D. Minton, D. Nesvorny et al. An Archean Heavy Bombardment from a Destabilized Extension of the Asteroid Belt. Nature 485: 78-81 (2012).

10. B. Fegley, K. Lodders. Very High Temperature Chemical Equilibrium Calculations with the CONDOR Code. 6th Ann. Meteoritical Soc. Meeting 5242. pdf. (2001).

11. P. Jenniskens, C.O. Laux, M.A. Wilson, E.L. Schaller. The Mass and Speed Dependence of Meteor Air Plasma Temperatures. Astrobiol. 4:81–94 (2004).

12. Hydrogen Production http://www1.eere.energy.gov/hydrogenandfuelcells/production/

13. G.J. Taylor. A New Moon for the Twenty-First Century. http://www.psrd. hawaii.edu/Aug00/newmoon.html

14. D.J. Lawrence, W.C. Feldman, B.L. Barraclough et.al. Global Elemental Maps of The Moon: The Lunar Prospector Gamma-Ray Spectrometer. Science 281: 1484 – 1489 (1998).

15. M.A. Wieczorek, R.J. Philllips. The "Procellarum KREEP Terrane": Implications for Mare Volcanism and Lunar Evolution. J. Geophys. Res. 105: 20417 – 20430 (2000).

16. M. Ohtake, T. Matsunaga, J. Haruyama et al. The Global Distribution of Pure Anorthosite on the Moon. Nature 461: 236 – 240 (2009).

17. R.J. Pike. Apparent Depth/Apparent Diameter Relation of Lunar Craters. Proc. Lunar Sci. Confer. 8th .pp. 3427-3436. (1977). http://cde.nwc.edu/SCI218/ course_documents/earth_moon/earth

18. G.J. Taylor. The Scientific Legacy of Apollo. Sci. Am. July 1994, 26–33.

19. P.H. Warren. "New" Lunar Meteorites: Implications for Composition of the Global Lunar Surface, Lunar Crust, and the Bulk Moon. Meteoritics Planet. Sci. 40: 477 – 506 (2005).

20. I.S. Kuvalaev. Inorganic Polyphosphate Functions at Various Stages of Cell Evolution. J. Biol. Phys. 20: 255 – 273 (2004).

21. B.C. Sales, L.A. Boatner, J.O. Ramey. Chromatographic Studies of the Structures of Amorphous Phosphates: A Review. J. Non – Cryst. Solids. 263: 155 – 166 (2000).

22. M.E. Jones, F. Lipmann. Chemical and Enzymatic Synthesis of Carbamyl Phosphate. Proc. Natl. Acad. Sci. U.S.A. 46: 1194 – 1205 (1960).

Multimedia Support

Youtube. "Solar System Evolution: The Nice Model". (1.28 min.).

Chapter 4. LIFE PRECURSORS FALL INTO PLACE

1. Geographical Review: Rock Types. http://geoinfo.amu.edu.pl/wpk/pe/a/harbbook/c_ii/chap02.html

2. SAAEA: Capturing and Converting CO_2 from Flue Gas Into Useful Materials. http://saaea.blogspot.com/2011/03

3. S.S. Goldich. A Study in Rock Weathering. J. Geol. 46:17 – 58 (1938).

4. M.D. Hopkins, T.M. Harrison, C.E. Manning. Constraints on Hadean Geodynamics from Mineral Inclusions in >4 Ga Zircons. Earth Planet. Sci. Lett. 298:367 – 376 (2010).

5. www.nasa.gov/mission.../fermi-thunderstorms.html

6. J. Oró, S.L. Miller, A. Lazcano. The Origin and Early Evolution of Life on Earth. Ann. Rev. Earth Planet Sci. 18:317-356 (1990).

7. S.A. Benner, H-J. Kim, M-J Kim, A. Ricardo. Planetary Organic Chemistry and the Origins of Biomolecules. Cold Spring Harb Persp. Biol. 2: a003467 (2009).

8. C. Darwin. Letter to J.D. Hooker, February 1, 1871.

9. E. Baldwin. An Introduction to Comparative Biochemistry. Cambridge University Press (1940).

10. A.B. McCallum. Paleochemistry of Body Fluids and Tissues. Physiol. Rev. 6:316 – 357 (1926).

11. F.H. Epstein. The Sea Within Us. J. Exp. Zool. 284:50-54 (1999).

12. M. Chaplin. Intracellular Water. www1.lsbu.ac.uk/water/cell.html

13. F. Robert, M. Chaussidon. A Paleotemperature Curve for the Precambrian Oceans Based on Silicon Isotopes in Cherts. Nature 443: 969-972 (2006).

14. E.A. Gaucher, S. Govindarajan, O.K. Ganech. Paleotemperature Trend for Precambrian Life Inferred from Resurrected Proteins. Nature. 451:704-707 (2008).

15. J.P. Knauth, D.A. Lowe. High Archean Climate Temperature Inferred from Oxygen Isotope Geochemistry of Cherts in the 3.5 Ga Swaziland Supergroup, South Africa. Geol. Soc. Am. Bull. 115: 566 – 580 (2003).

Chapter 5. COMPLEX MOLECULES START TO FORM

1. J.B.S. Haldane. The Origin of Life. The Rationalist Annual. 148:3-10 (1929).

2. A.G. Griesbeck, J. Mattay. Synthetic Organic Photochemistry. M Dekker, N.Y. (2005).

3. A. Shimoyama, H. Naraoka, M. Komiya, K. Harada. Analysis of Carboxylic Acids and Hydrocarbons in Antartic Carbonaceous Chondrites, Yamato-74662 and Yamato-793321. Geochem. J. 23:181-193 (1989).

4. S. Pizzarello, G.W.Cooper. G.J.Flynn. The Nature and Distribution of the Organic Material in Carbonaceous Chondrites and Interplanetary Dust Particles. In Meteorites and the Early Solar System II. D. Lauretta, L.A. Leshin, H.Y. Mc Sween Eds.Univ. of Arizona Press, pp. 625-651 (2006).

5. Z. Martins, J.S. Watson, M.A. Sephton, et al. Free Dicarboxylic and Aromatic Acids in the Carbonaceous Chondrites Murchison and Orgueil. Meteor. Planet. Sci. 41:1073-1080 (2006).

6. H. Schulz. Short History and Present Trends of Fischer-Tropsch Synthesis. Appl. Catalysis A: General. 186:3-12 (1999).

6a. M.P. Powner, J.D. Sutherland, Prebiotic Chemistry: A New Modus Operandi. Phil. Trans. R. Soc. B. 366:2870-2877 (2011).

7. J.A. Baross, S.E. Hoffman. Submarine hydrothermal vents and Associated Gradient Environments as Sites for the Origin and Evolution of Life. Orig. Life Evol. Biosph. 15:327 – 345 (1985).

8. E.T. Parker, H.J. Cleaves, J.P. Dworkin et al. Primordial Synthesis of Amines and Amino Acids in a 1958 Miller H_2S-Rich Spark Discharge Experiment. Proc. Natl. Acad. Sci. U.S.A. 108:5526-5531 (2011).

9. J. Eichberg, E.Sherwood, D.E. Epps, J. Oró. Cyanamide Mediated Synthesis under Plausible Primitive Earth Conditions. IV. The Synthesis of Acylglycerols. J. Mol. Evol. 10:221-230 (1977).

10. M. Rao, M.R. Eichberg, J. Oró. Synthesis of Phosphatidylcholine under Possible Primitive Earth Conditions. J. Mol. Evol. 18: 196-202 (1982).

11. M. Rao, M.R. Eichberg, J. Oró. Synthesis of Phosphatidylethanolamine under Possible Primitive Earth Conditions. J. Mol. Evol. 25:1-6 (1987).

12. S. Baoukina, L. Monticelli, H. J. Risselada, et. al. The Molecular Mechanism of Lipid Monolayer Collapse. Proc. Natl. Acad. Sci. U.S.A. 105:10803-10808 (2008).

13. D. Segré, D. Ben-Eli, D. Deamer, D. Lancet. The Lipid World. Orig. Life Evol. Biosph. 31:119-145 (2001).

14. M. De Rosa, A. Gambacorta. Lipid Biogenesis in Archaebacteria System. Appl. Microbiol. 7:278-285 (1986).

15. S.W. Fox, K. Harada. The Thermal copolymerization of Amino Acids Common to Protein. J. Am. Chem. Soc. 82: 3745-3751 (1960).

16. S.W. Fox, J.R. Jungck, T. Nakashima. From Proteinoid Microsphere to Contemporary Cell: Formation of Internucleotide and Peptide Bonds by Proteinoid Particles. Orig. Life 5:227-237 (1974).

17. C.L. Apel, D.W. Deamer, M.N. Mautner. Self-assembled Vesicles of Monocarboxylic Acids and Alcohols: Conditions for Stability and for the Encapsulation of Biopolymers. Biochim. Biophys. Acta 1559:1-9 (2002).

18. P-A. Monnard, D.W. Deamer. Membrane Self- Assembly Processes: Steps Towards The First Cellular Life. Anat. Rec. 268:196-207 (2002).

19. A. Butlerow. Formation Synthétique d'une Substance Sucrée : Compt. Rend. Acad. Sci. 53:145-147 (1861).

20. R. Breslow. On the Mechanism of the Formose Reaction. Tetrahedr. Lett. 21:22-26 (1959).

21. B. Moore, T. A. Webster. Action of Light Rays on Organic Compounds, and the Photosynthesis of Organic from Inorganic Compounds in Presence of Inorganic Colloids. Proc. R. Soc. Lond. B. 90:168-186 (1918).

22. J.C. Irvine, G.V. Francis. On Examination of Photosynthetic Sugars by the Methylation Method. J. Ind. Eng. Chem. 16:1019 (1924).

23. E.C.C. Baly, J.B. Davies, M.R. Johnson. H. Shanassy. The Photosynthesis of Naturally Occurring Compounds. I. The Action of Ultra-Violet Light on Carbonic Acid. Proc. R. Soc. Lond. A. 116:197-211 (1927).

24. E.C.C. Baly, I.M. Heilbron, D.P. Hudson. CXXX. Photocatalysis. Part II. The Photosynthesis of Nitrogen Compounds from Nitrates and Carbon Dioxide. Trans. Chem. Soc. 121:1078-1088 (1922).

25. R.F. Socha, A.H. Weiss, M.M. Sakharov. Homogenously Catalyzed Condensation of Formaldehyde to Carbohydrates. J. Catal. 67:207-217 (1981).

26. A.W. Schwartz, R.M. de Graaf. Prebiotic Synthesis of Carbohydrates. A Reassessment. J. Mol. Evol. 36:101-106 (1993).

27. O. Pestunova, A. Simonov, V. Snytnikov, et. al. Putative Mechanism for the Sugar Formation on Prebiotic Earth Initiated by UV-Radiation. Adv. Space Res. 36:214-219 (2005).

28. K.J. Zahnle. Earth's Earliest Atmosphere. Elements 2:217-222 (2006).

29. R. Larralde, M.P. Robertson, S.L. Miller. Rates of Decomposition of Ribose and other Sugars : Implications for Chemical Evolution. Proc. Natl. Acad. Sci. U.S.A. 92:8158-8160 (1995).

30. J.B. Lambert, S.A. Gurusamy-Thangevelu, K.Ma. The Silicate-Mediated Formose Reaction: Bottom-up Synthesis of Sugar Silicates. Science 327:984-986 (2010).

31. R. Pfanstiel, R.K. Iler. Potassium Metaphosphate: Molecular Weight, Viscosity Behavior and Rate of Hydrolysis of Non-crossed-linked Polymer. J. Am. Chem. Soc. 74:6059-6064 (1952).

32. O.P. Pestunova, A.N. Simonov, V.N. Snytnikov, V.N. Parmon. Prebiotic Carbohydrates and their Derivatives. In Biosphere, Origin and Evolution. N. Dobretsov et al. (eds.) Springer, Berlin (2008).

33. R.B. Stockbridge, C.A. Lewis, Y. Yuan, R. Wolfenden. Impact of Temperature on the Time Required for the Establishment of Primordial Biochemistry, and for the Evolution of Enzymes. Proc. Natl. Acad. Sci. U.S.A. doi: 10.1073/pnas. 1013647107, (Dec. 1, 2010).

Chapter 6. INFORMATIONAL MACROMOLECULES AND THE EMERGENCE OF CELLS

1. K.J. Zahnle. The Earth's Earliest Atmosphere. Elements 2:217-222 (2006).

2. D. Segré, D. Ben-Eli, D.W.Deamer et al. The Lipid World. Orig. Life Evol. Biosph. 31:119-145 (2001).

3. E. Budin, J.W. Szostak. Expanding Roles for Diverse Physical Phenomena During the Origin of Life. Ann. Rev. Biophys. 39:245-263 (2010).

4. U.J. Meierhenrich, J.-J. Filippi, C. Meinert et al. On the Origin of Primitive Cells: From Nutrient Intake to Elongation of Encapsulated Nucleotides. Angew. Chem. 49:3738-3750 (2010).

5. C. Woose. The Universal Ancestor. Proc. Natl. Acad. Sci. U.S.A. 95:6854-6859 (1998).

6. K. Kruger, P.J. Grabowski, A.J. Zang et al. Self-Splicing RNA, Autoexcision and Autocyclization of the Ribosomal RNA Intervening Sequence of Tetrahymena. Cell 31:147-157 (1982).

7. C. Guerrier-Takada, K. Gardiner, I. Tinoco et al. The RNA Moiety of Ribonuclease P is the Catalytic Subunit of the Enzyme. Cell 35:849-857 (1983).

8. M. Eigen. The Origin of Genetic Information: Viruses as Models. Gene 135:37-47 (1993).

9. M. Eigen. J. McCaskill, P. Schuster. The Molecular Quasi-Species. Adv. Chem. Phys. 75:149-263 (1989).

10. F.H.C. Crick. On The Origin of the Genetic Code. J. Mol. Biol. 38:367-379 (1968).

11. F.H.C. Crick, S. Brenner, A. Klug et al. A Speculation on the Origin of the Genetic Code. Orig. Life 7:389-397 (1976).

12. H.F. Noller, V. Hoffarth, L. Zimniak. Unusual Resistance of Peptidyl Transferase to Protein Extraction Procedures. Science 256:1416-1419 (1992).

13. R. Schekman, A. Weiner, A. Kornberg. Multienzyme Systems for DNA Replication. Science 186:987-993 (1974).

14. C.W. Greider, E.H. Blackburn. The Telomere Terminal Transferase of Tetrahymena is a Ribonucleoprotein Enzyme with Two Kinds of Primer Specificity. Cell 51:887-898 (1987).

15. M.J. Curcio, K.M. Derbyshire. The Outs and Ins of Transposition: From Mu to Kangaroo. Nature Rev. Mol. Cell Biol. 4:865-877 (2003).

16. Y. Xiong, T.H. Eickbush. Origin and Evolution of Retroelements Based Upon Their Reverse Transcriptase Sequences. EMBO J. 9:3353-3362 (1990).

17. A. Lazcano, V. Valverde, G. Hernández et al. On the Early Emergence of Reverse Transcription: Theoretical Basis and Experimental Evidence. J. Mol. Evol. 35:524-536 (1992).

18. P. Schuster. How Does Complexity Arise in Evolution? Complexity 2:22-30 (1996).

19. J.W. Schopf. Microfossils of the Early Archean Apex Chert-New Evidence of the Antiquity of Life. Science 260:640-646 (1993).

20. J.M. García-Ruiz, S.T. Hyde, A.M. Carnerup, A.G. Christy, et al. Self-Assembled Silica-Carbonate Structures and Detection of Ancient Microfossils. Science 302:1194-1197 (2003).

21. Y. Ueno. Y. Isozaki, H. Yurimoto et al. Carbon Isotopic Signatures of Individual Archean Microfossils (?) from Western Australia. Int. Rev. Geol. 43:196-212 (2001).

22. J.W. Schopf, A.B. Kudryavtsev, D.G. Agresti et al. Laser-Raman Imagery of Earth's Earliest Fossils. Nature 416:73-76 (2002).

23. S. Derenne, F. Robert. A. Skrzypczak-Bonduelle, D.Gourier et al. Molecular Evidence for Life in the 3.5 Billion Year Old Warrawoona Chert. Earth Planet. Sci. Lett. 272:476-480 (2008).

24. M. Eigen, B.F. Lindemann, M. Tietze et al. How Old is The Genetic Code? Statistical Geometry of tRNA Provides an Answer. Science 244:673-679 (1989).

25. C. de Duve. Vital Dust. Basic Books, N. York, 1994.

26. M. Eigen. Steps Towards Life. Oxford University Press, 1992, p. 31.

27. C. Darwin. Letter to G.C. Wallich, 28 March, 1882. Letter 13747.

Chapter 7. CHANCE, SPONTANEITY AND INFORMATION

1. E. Schrödinger. What is Life? Cambridge University Press. 1967.
2. H. Atlan. L'Organisation Biologique et la Théorie de L'Information. Hermann, Paris 1972, p. 175.
3. R. Schoenheimer. The Dynamic State of Body Constituents. Harvard University Press, 1942.
4. H.R Maturana, F.J. Varela. Autopoiesis and Cognition. Reidel Publ. Co. Dordrecht, 1980.
5. H. Atlan. L'Organisation Biologique et la Théorie de L'Information. Hermann, Paris, 1972, pp. 66-73.

Chapter 8. THE ORIGIN OF LIFE AND DARWIN'S CONCLUSION

1. I. Prigogine, I. Stenger. Order out of Chaos. Flamingo. London, 1985, Chapter 9.

2. P.C.W. Davies. The Physics of Time Asymmetry. Univ. of California Press, Berkeley 1974, 153 pp.

3. R. Highfield, P. Coveney. The Arrow of Time. W.H. Allen 1990.

4. J.D.Barrow, F.J. Tipler. The Anthropic Cosmological Principle. Oxford University Press, 1986, p. 653.

5. A. Pais. Subtle is the Lord. The Science and the Life of Albert Einstein. Oxford University Press, 1982. p. 177-265.

6. A. Friedmann. Über die Krümmung des Raumes. Z. Physik 10:377-386 (1922).

7. H. Kragh. Cosmology and Controversy. Princeton University Press, 1999. 501 pp.

8. A. Calaprice, T. Lipscombe. Albert Einstein, A Biography. Greenwood Press, Westport CT, 2005.p.72.

9. A.G. Riess, A.V. Filippenko, P. Challis et al. Observational Evidence from Supernovae for an Accelerating Universe and a Cosmological Constant. Astron. J. 116:1009-1038 (1998).

10. S. Perlmutter, G. Aldering, G. Goldhaber et al. Measurements of Omega and Lambda from 42 High-Redshift Supernovae. Astrophys. J. 517:565-586 (1999).

11. E. Hubble. Extragalactic Nebulae. Astrophys. J. 64:321-369 (1926).

12. B. Pascal. Pensées, 1671, N°72.

13. Saint Augustine. Sermo 1,5.

14. L. Wittgenstein. Tractatus Logico-Philosophicus, 6.4312.

15. T. Aquinas. Sum. Theol. 1a, q.45, a.5.

16. D. Atkatz. H. Pagels. The Origin of The Universe as a Quantum Tunneling Event. Phys. Rev. D. 25:2065-2073 (1982).

17. J. Gleick. Chaos: The Making of a New Science. Viking. 1982.

18. E.N. Lorenz. Atmospheric Predictability Experiments with a Large Numerical Model. Tellus 34:505-513 (1982).

19. C. Darwin. The Origin of Species. Penguin Books, Lond. 1982. p. 458.

SUBJECT INDEX

www.ingramcontent.com/pod-product-compliance
Lightning Source LLC
Chambersburg PA
CBHW040826180526
45159CB00001B/76